A GUIDE TO
SKYWATCHING

ATLAS COELESTIS;
seu
HARMONIA
MACROCOSMICA.

A Guide to
SKYWATCHING

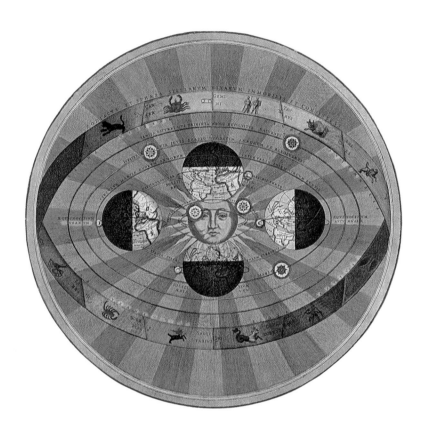

David H. Levy

CONSULTANT EDITOR
Dr John O'Byrne

FOG CITY PRESS

Published by Fog City Press
814 Montgomery Street
San Francisco, CA 94133 USA

Copyright 1994 © US Weldon Owen Inc.
Copyright 1994 © Weldon Owen Pty Limited
This edition 2002
Reprinted 2002, 2003

CHIEF EXECUTIVE OFFICER John Owen
PRESIDENT Terry Newell
PUBLISHER Lynn Humphries
MANAGING EDITOR Janine Flew
ART DIRECTOR Kylie Mulquin
EDITORIAL COORDINATOR Tracey Gibson
EDITORIAL ASSISTANT Kiren Thandi
PRODUCTION MANAGER Caroline Webber
PRODUCTION COORDINATOR James Blackman
BUSINESS MANAGER Emily Jahn
VICE PRESIDENT INTERNATIONAL SALES Stuart Laurence
EUROPEAN SALES DIRECTOR Vanessa Mori

PROJECT EDITOR Lu Sierra
COPY EDITOR Gillian Hewitt
DESIGNER Clare Forte
DESIGN ASSISTANT Stephanie Cannon
JACKET DESIGN John Bull
PICTURE RESEARCH Gillian Manning
ILLUSTRATIONS Steven Bray, Lynette R. Cook, David Wood
STARFINDER CHARTS Wil Tirion

ISBN 1 877019 06 2

Color reproduction by Colourscan Co Pte Ltd
Printed by Kyodo Printing Co (S'pore) Pte Ltd
Printed in Singapore

A Weldon Owen Production

Look at the stars! Look, look up at the skies!
O look at all the fire-folk sitting in the air!
The bright boroughs, the circle-citadels there!

The Starlight Night,
GERARD MANLEY HOPKINS (1844–1889), English poet

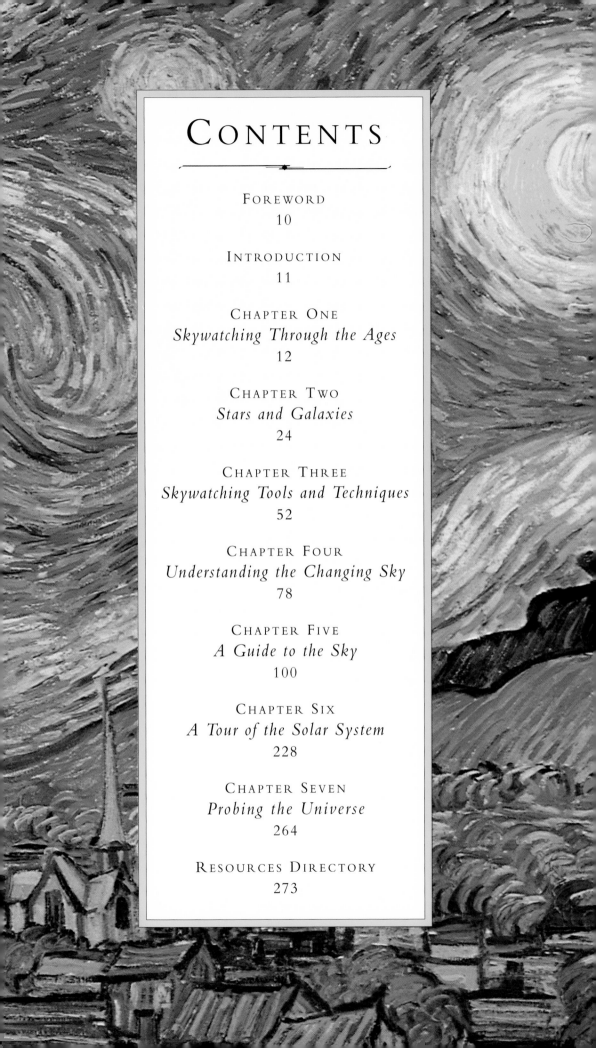

CONTENTS

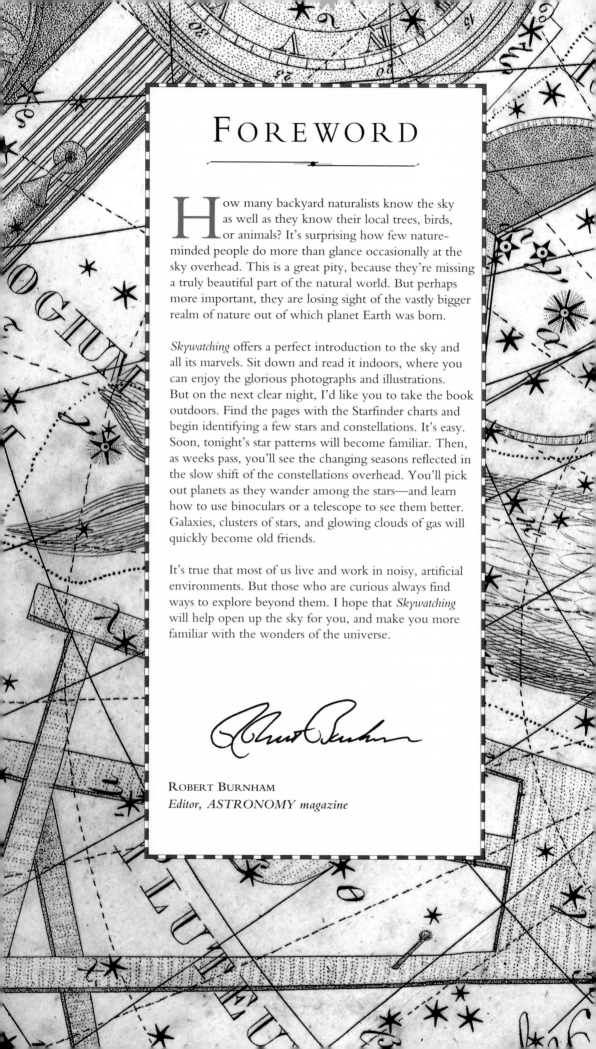

FOREWORD

How many backyard naturalists know the sky as well as they know their local trees, birds, or animals? It's surprising how few nature-minded people do more than glance occasionally at the sky overhead. This is a great pity, because they're missing a truly beautiful part of the natural world. But perhaps more important, they are losing sight of the vastly bigger realm of nature out of which planet Earth was born.

Skywatching offers a perfect introduction to the sky and all its marvels. Sit down and read it indoors, where you can enjoy the glorious photographs and illustrations. But on the next clear night, I'd like you to take the book outdoors. Find the pages with the Starfinder charts and begin identifying a few stars and constellations. It's easy. Soon, tonight's star patterns will become familiar. Then, as weeks pass, you'll see the changing seasons reflected in the slow shift of the constellations overhead. You'll pick out planets as they wander among the stars—and learn how to use binoculars or a telescope to see them better. Galaxies, clusters of stars, and glowing clouds of gas will quickly become old friends.

It's true that most of us live and work in noisy, artificial environments. But those who are curious always find ways to explore beyond them. I hope that *Skywatching* will help open up the sky for you, and make you more familiar with the wonders of the universe.

ROBERT BURNHAM
Editor, ASTRONOMY magazine

INTRODUCTION

Since ancient times, human beings have been fascinated by the origins and movements of the stars and planets. Astronomy began with the first humans ever to look at the night sky and notice the stars. About six thousand years ago, astronomer-astrologers appeared—priests whose task it was to track the heavens to learn the gods' intent. Today, we rely on space missions and observatories to explain and record what our ancestors could only wonder at.

What modern astronomers have discovered is a world of ceaseless activity. While we work, play, and sleep, the universe is constantly expanding; old stars are dying in cataclysmic fireballs; and new stars are being born. On a dark night in the countryside, far from the glare of city lights, gaze upward and you will see the beautiful spectacle of thousands of stars belonging to our galaxy, the Milky Way. Look with a telescope and you may see glittering star clusters, wispy nebulae, misty galaxies, and intriguing double stars. There are boundless riches here. No wonder humans have been looking skyward for millennia, wondering how far away the stars were, what they were made of, and how they got there. Science and technology may have resolved some of the sky's mysteries, but still millions more remain to be explored.

This guide is designed to increase your awareness and enjoyment of the sky above you. Combining facts compiled by experts with the work of talented photographers and artists, it aims to take the world of skywatching out of dusty libraries and make it a part of people's daily lives. Whether your interests lie in armchair astronomy, practical skywatching, astral photography, or any other aspect of astronomy, this book will guide you on a journey into the countless wonders of the universe.

The Editors

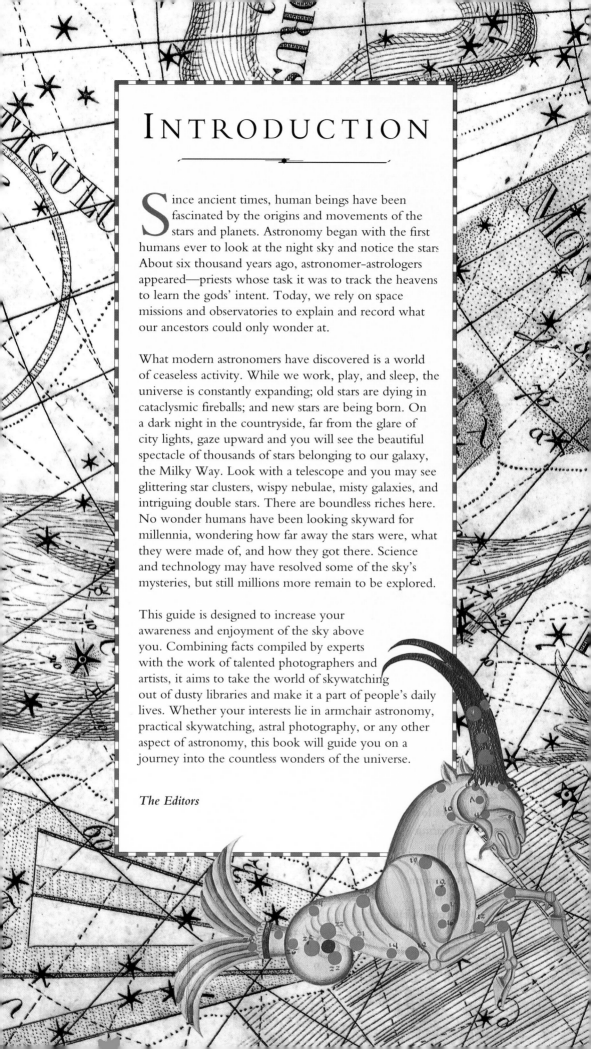

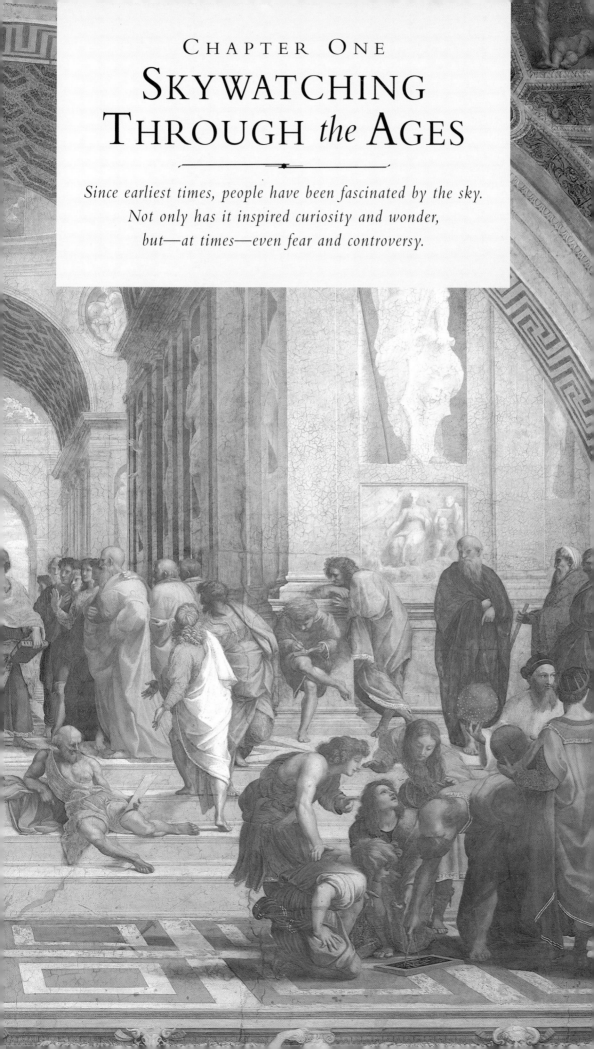

CHAPTER ONE
SKYWATCHING
THROUGH *the* AGES

Since earliest times, people have been fascinated by the sky.
Not only has it inspired curiosity and wonder,
but—at times—even fear and controversy.

THE FIRST SKYWATCHERS

From the time we first connected the motions of
the heavens with the changing of day into night,
we have been fascinated by astronomy.

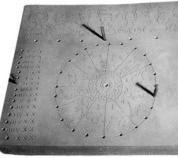

BABYLONIAN TABLET *recording astronomical information, dating from around 550 BC*

THE ROMAN CALENDAR *featured a 7 day week and 12 months, each of around 30 days. Days, weeks, and months can be counted off using the holes in this stone block.*

I n early times, skywatchers saw the heavens in religious terms. The sky was the home of the gods as they controlled day and night, storms, and the great eclipses of the Sun and Moon. In many cultures, personalities were attributed to celestial patterns and objects and were thought to influence people's everyday lives. Astronomer-priests studied the sky, kept records, compiled calendars, and acted as custodians of the legends relating to the sky. There was no clear distinction between astronomy and astrology.

STONEHENGE, *in southwestern England, is one of Europe's most striking megalithic monuments. Initial building has been dated at about 1800 BC, and today's stone circles were constructed in several stages over the next 400 years. It is aligned to receive the rays of the mid-summer sunrise, and may have been used to predict the motions of the Sun and the Moon, including their eclipses.*

THE BABYLONIANS

Among the first people known to have kept astronomical records were the Akkadians who lived some 4,500 years ago in the northern part of what later became Babylonia. There is some evidence that their ideas about the motions of the Sun, the Moon, and the planets were later codified by the Babylonians. From their observations, the astronomer-priests of Babylonia could predict the courses of wandering objects in the sky.

THE FIRST CALENDARS

With a calendar that dates back to 1300 BC, the Chinese are believed to be the earliest calendar makers. The Babylonians and the ancient Egyptians also developed accurate calendars from their studies of the heavens.

Having a calendar meant people had a record of the seasons and thus knew when to plant and harvest their crops. For the Egyptians, whose economy depended on agriculture, a calendar meant

NUT, *the Egyptian goddess of the sky, is usually depicted arched over the reclining Earth god, Geb, her husband. The Sun god, Ra, was thought to travel across the sky each day in his boat.*

NAVIGATION *by the stars requires precise knowledge of star positions and motions. Here, a medieval navigator is making his measurements on shore.*

that they could predict when the Nile's annual flood would irrigate their fields. The priests would wait for the morning that Sirius, the brightest star in the sky, first appeared after being blocked by the Sun. They would then use this "heliacal rising" to predict the annual floods.

From earliest times, there has been a 7 day week, to match each quarter phase of the Moon, and the 12 months of the year reflect the Moon's completion of its cycle of phases 12 times a year.

CHINESE ASTRONOMERS

The Chinese have a long history of dedicated sky-watching, and there is evidence that they recorded a close grouping of the bright planets in about 2500 BC.

In the fourth century BC they produced the earliest

known atlas of comets, the *Book of Silk*. A silk ribbon about 5 feet (1.5 m) long, it illustrates 29 forms of comets and lists the various types of catastrophes they heralded. The work was discovered in a tomb in 1973.

NAVIGATION

Ever since they first took to the water in boats, sailors have had a close relationship with the heavens. Out of sight of land, they used the positions of the stars to guide them. Polynesian islanders knew how to navigate across the vast stretches of the Pacific between, for example, Tahiti and Hawaii, by plotting their course by the stars. They learned the positions of the stars and the prevailing wind patterns through poetry that they learnt by heart, passed down from generation to generation.

NAVIGATIONAL INSTRUMENT
of a kind used for thousands of years by the inhabitants of the Marshall Islands in the Pacific

STAR CLOAK *worn by a Pawnee Indian and photographed in the last century. The Pawnee regarded some of the stars as gods and sought their favors through elaborate rituals.*

SHAPES *in the* SKY

From the dawn of time, skywatchers have projected images of their own invention onto the stars, filling the heavens with gods, animals, and fantastic creatures.

From earliest times, people have grouped stars into constellations, turning the outlines they formed into figures and animals relating to local customs and creating elaborate stories around them. Sometimes, of course, they were thought of as deities; sometimes they were seen as something more down-to-earth—for instance, the stars of Corvus, the Crow, were seen by the Arabs as a tent. Some star groupings are so striking that throughout the world there are myths associated with them.

A good example is the constellation of Ursa Major, the Great Bear, which is said to be the Callisto of Greek legend, who was transformed by Zeus or, according to another account, his wife Hera, into a bear. This constellation was also seen as a bear by many civilizations, including several North American Indian tribes. It contains seven bright stars, best known as the Big Dipper or the Plough, and its distinctive shape has been noted by poets such as Homer,

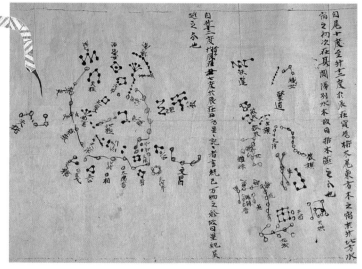

A SNAKE *such as this (above left), from an Australian Aboriginal bark painting, is commonly seen in star patterns.*

CHINESE STAR MAP *This star map from Tunhuang, China, dates from around* AD *940. Except for the simple star maps on astrolabes, this is the oldest known portable star map. Ursa Major is plainly visible near the bottom of the left half of the map.*

Shakespeare, and Tennyson. In Hindu mythology these seven stars represent the homes of the seven great sages; the Egyptians saw them as the thigh of a bull; the ancient Chinese saw them as the masters of the reality of heavenly influences; and Europeans saw them forming a wagon, the Anglo-Saxons associating the wagon with the legendary King Arthur.

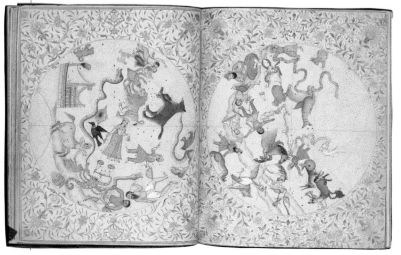

INDIAN STAR CHART *This beautiful rendition of the constellations, featuring ancient Indian and Islamic patterns, accompanied a horoscope commissioned by an Indian monarch for his son in 1840.*

ZODIACAL SIGNS *(below) ring the world, which is composed of four basic elements—earth, air, fire, and water— in this encyclopedia dating from the fifteenth century.*

MAORI SKYLORE *Depicted in this tribal carving (above) is the separation of Ranginui (the sky father) and Papatuanuku (the Earth mother) by their sons who, living in darkness, longed to experience light and day (represented by the spirals between the figures).*

PEGASUS IN A NEW FORM

The constellation of Pegasus is supposed to represent a winged horse, but to most people it looks more like a square, so it is generally known as the Great Square. But since it is so prominent in the evening sky during fall in North America, why not call it Pegasus, the Baseball Diamond? There are stars for each of the four bases, one each for pitcher, batter, and catcher, and the nearby Milky Way can represent the fans in the stadium!

THE ZODIAC AND ASTROLOGY

The zodiac is a belt of 12 constellations that straddle the sky.

Largely within this belt lie the apparent paths of the Sun, the Moon, and the bright planets. The zodiacal signs are the most ancient of the 88 constellations recognized today and act as markers in astrology —the belief that the stars and planets influence human affairs.

Although there is no scientific evidence that the way the planets line up has ever affected anyone, at one time it would have been commonplace for a ruler to consult his astrologer before going into battle, or a businessman to check whether the planets favored a particular deal. Most people today are interested in astrology just for fun, but quite a few still take it seriously. Just remember that if you are worried about the gravitational pull that was exerted by Mars on your character when you were born, you can be sure that the pull exerted by your obstetrician was considerably stronger!

THE TWELVE APOSTLES *replace the traditional zodiacal constellations in this star map produced by Julius Schiller, a devout Catholic, for his 1627 book* Coelum Stellatum Christianum.

CHARTING *the* UNIVERSE

Ptolemaic theory placed the Earth at the center of the universe, and it took until the sixteenth century for Copernicus to dethrone Earth and prove that Aristarchus had been right all along.

CLAUDIUS PTOLEMAEUS, *better known as Ptolemy, was a brilliant and prolific author on a diverse range of topics, and made the first truly scientific maps of the Earth and heavens.*

The ancient Greeks were the first people to try to explain why natural events occurred without reference to supernatural causes and, in so doing, they changed astronomy from a mystical cult into a science. Greek thinkers came to realize that the prevailing astrological beliefs did not correspond with the "rules" of the universe that they were beginning to uncover.

Thales, who lived in the sixth century BC, was the first of the great Greek philosophers and a keen traveler. He brought back to Greece the knowledge and records of the Babylonians and the Egyptians and put forward theories that

ARISTOTLE, *whose theory of an Earth-centered universe dominated thinking for 1,800 years, is portrayed in this painting by Rembrandt (1606–69) contemplating a bust of the Greek poet Homer.*

HIPPARCHUS *(right), a Greek astronomer and mathematician of the second century BC, produced the first known star catalogue and discovered the precession of the equinoxes (see p. 87).*

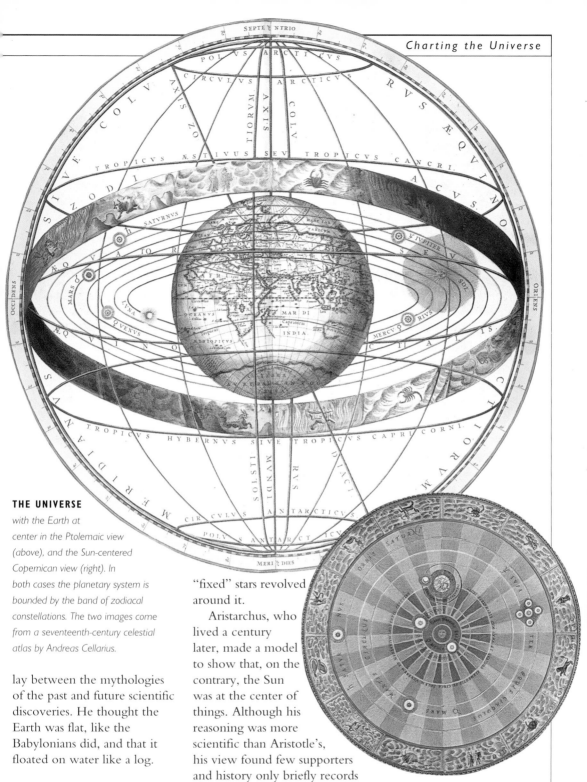

THE UNIVERSE

with the Earth at center in the Ptolemaic view (above), and the Sun-centered Copernican view (right). In both cases the planetary system is bounded by the band of zodiacal constellations. The two images come from a seventeenth-century celestial atlas by Andreas Cellarius.

lay between the mythologies of the past and future scientific discoveries. He thought the Earth was flat, like the Babylonians did, and that it floated on water like a log.

CONFLICTING THEORIES

Aristotle, who lived from 384 to 322 BC, was one of the most influential of all the Greek philosophers, and he put forward three experimental proofs to explain that the Earth was round. He also thought that the Earth formed the center of the universe, and that the Sun, the Moon, the planets, and a sphere containing all the "fixed" stars revolved around it.

Aristarchus, who lived a century later, made a model to show that, on the contrary, the Sun was at the center of things. Although his reasoning was more scientific than Aristotle's, his view found few supporters and history only briefly records his heliocentric ideas.

PTOLEMY

Ptolemy of Alexandria, another gifted Greek astronomer and scholar, published his *Almagest* in about AD 140. In this remarkable encyclopedia of ancient science he used centuries of Babylonian observations of the motions of the planets to buttress his argument for an Earth-centered

universe. His intricate system of circles within circles proved to be a marvelous mathematical method for predicting the motions of the planets.

Ptolemy's "system of the world", known as the Ptolemaic system, embodied ideas that were to rule the world of astronomy for about 1,500 years. His death marks the end of the classical era of astronomy.

19

COPERNICUS

The process of breaking down Ptolemy's system began with the work of a Polish churchman, Nicolaus Copernicus, who was born in 1473. From his early years, Copernicus thought that the Sun lay at the center of the system of planets and stars, rather than the Earth, but he did not complete his formal work on the subject until he was an old man.

In 1543, just before his death, he published his masterpiece *On the Revolution of the Celestial Spheres*, which was to change humanity's view of the cosmos. Not surprisingly, it met with great hostility from the Church, which held that God had created a universe with the Earth at its heart.

As a system of mathematical prediction, the Copernican view was no more successful than the Ptolemaic, but two other events made the Copernican revolution

NICOLAUS COPERNICUS *(1473–1543) is the Latin name by which the Polish astronomer Niklas Koppernick is generally known. His position as Canon of the cathedral at Frauenberg allowed him to pursue his studies of mathematics, astronomy, medicine, and theology.*

inevitable: Tycho Brahe's amazingly precise observations of the sky, and Galileo's use of a simple spyglass.

TYCHO BRAHE

One evening in 1572, Danish astronomer Tycho Brahe saw a brilliant new star on the constellation of Cassiopeia. He was so surprised, it is said, that he asked his neighbor to hit him to make sure he wasn't dreaming. We now know that this new star was a supernova —the explosion of a dying star that is so violent that it outshines the combined light of all the stars in our galaxy. In 1604, a second bright supernova blazed forth in the sky. These discoveries shattered a cornerstone of Ptolemaic thinking:

that the outermost sphere, containing all the stars, was unchanging. It was as though the heavens themselves had joined the Renaissance in Europe.

KEPLER

Tycho's other major contribution was to pass on to his assistant, Johannes Kepler, the records of his observations

SUPERNOVA OF 1572 *is marked as 'I' in this engraving from Tycho Brahe's book* De Stella Nova.

URANIBORG, *Tycho Brahe's (1546–1601) observatory on the Danish island of Hven from where he made the precise observations that were his greatest contribution to astronomy.*

of the motions of the planets that he had made from 1576 to 1597. These represent a pinnacle of achievement of naked-eye astronomy. Kepler took this data and labored for years to produce his three laws of planetary motion, laws that enabled him to predict the positions of the planets more effectively than Ptolemy or any of his successors.

GALILEO AND THE TELESCOPE

In 1609, an Italian scientist named Galileo Galilei heard of an astonishing invention: by looking through two glass lenses held at a fixed distance from each other and the eye, distant objects could be magnified. Making a simple telescope for himself, he turned it toward the sky, and among his many discoveries he found that the giant planet Jupiter had four moons clearly revolving about it.

Galileo's telescope revealed a miniature version of Copernicus's Solar System, with the moons circling Jupiter in simple, nearly circular orbits.

JOHANNES KEPLER *(1571–1630) joined Tycho Brahe's research staff in Prague and succeeded him as Imperial Mathematician when Tycho died.*

When Galileo published his discoveries it was clear that he had abandoned the Ptolemaic system, but it was not until 1616 that the Church quietly warned him to change his views. In 1632, however, he published his book *Dialogue Concerning the Two Great World Systems*, in which three characters discuss the nature of the universe. One of them (with the suggestive name of Simplicio) supports Ptolemaic theory. The Pope, believing that Galileo was ridiculing him, put Galileo at the mercy of the Holy Office of the Inquisition and he was accused of heresy.

Forced to "abandon the false opinion that the Sun is the center of the world," Galileo lived out his days under house arrest. But the Inquisition could do nothing to stem the influence of his discoveries, which changed the face of astronomy. The

Catholic Church reconsidered the case and absolved Galileo from any wrongdoing in 1992. In 1989, a spacecraft named after him was launched to study Jupiter and its satellites—the moons that he was the first to see.

GALILEO'S *first telescopes were not as effective as an inexpensive telescope of today, but they changed our view of the universe.*

GALILEO'S TRIAL *in Rome in 1633 was marked by special indulgence and he was never jailed. A guilty verdict still resulted for having "held and taught" the Copernican doctrine.*

SKYWATCHING TODAY

Skywatchers today benefit from the dramatic developments in astronomy that have taken place since the turn of the century.

THE KECK TELESCOPE, *on Mauna Kea's summit, is the world's largest.*

Most skywatchers today see their observing as a casual activity—something they do just for pleasure. Amateur astronomers take it a little more seriously, perhaps recording their achievements in a logbook.

While it could be said that professional astronomers are distinguished by the equipment they use, this distinction is becoming a little blurred now that amateurs have access to powerful computers and electronic cameras.

An enormous amount of fundamental work is done by both amateur and professional astronomers to form the basis of the occasional major new discovery which makes the news. These discoveries highlight the explosive growth in our understanding of the universe that has taken place this century.

Ever-larger telescopes have played a major part in these developments, as have more sophisticated detectors for observing the sky. At the turn of the century, astronomers were just becoming accustomed to recording their observations on photographic plates. Now,

SOME MILESTONES IN TWENTIETH CENTURY ASTRONOMY

1900-1919

1905
Einstein's special theory of relativity published.

1908
Hertzsprung describes the giant and dwarf divisions of stellar types.

1912
Leavitt describes the period-luminosity law for Cepheid variables.

1916
Publication of Einstein's general theory of relativity, which predicts the expanding universe.

1917
100 inch (2.5 m) reflector at Mount Wilson, California, completed.

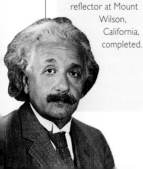

1920-1939

1923
Hubble shows that the spiral nebulae are galaxies lying outside the Milky Way.

1929
Hubble presents evidence for the expansion of the universe.

1930
Discovery of Pluto by Tombaugh.

1931
Jansky discovers radio waves from space.

1937
Reber reveals radio waves coming from Milky Way, using his radio telescope.

1938
Theory of stellar energy using nuclear reactions announced by Hans Bethe.

1940-1949

1942
Hey and colleagues detect radio waves from the Sun.

1946
Hey, Parsons, and Phillips identify the most powerful radio source—Cygnus A.

1948
200 inch (5 m) reflector at Mount Palomar, California, completed.

Bond and Gold propose the steady-state cosmological theory.

Gamow and Alpher describe the origin of the elements in the Big Bang.

1950-1959

1952
Baade announces a doubling of the distance-scale for the galaxies.

1957
Sputnik I launch marks the beginning of the space age.

1959
First pictures of the Moon's far side taken by space probes.

1960-1969

1961
Gagarin makes first manned space flight.

1963
Identification of quasars by Schmidt at Mt Palomar, California.

1965
Discovery of the 3°K microwave background radiation by Penzias and Wilson.

1967
Discovery of pulsars, by Bell-Burnell at Cambridge, England.

1969
Moon landing: Armstrong and Aldrin, Apollo 11.

all the world's major observatories use electronic detectors like CCDs (charge-coupled devices), although photography still has its place.

In 1900, the world's largest operating telescope was the 40 inch (1 m) diameter refractor at Yerkes Observatory at Williams Bay, Wisconsin. By 1917, a 60 inch (1.5 m) reflector was in operation at Mount Wilson, California; a 72 inch (1.8 m) one had been built in Victoria, British Columbia; and a mighty 100 inch (2.5 m) telescope had been completed at Mount Wilson, all of them peering deeper into the sky. In 1948, the enormous 200 inch (5 m) Hale telescope opened at Mount Palomar, California.

The 1960s and '70s saw a proliferation of large telescopes, especially in the Southern Hemisphere, and the setting up of large numbers of radio telescopes. In the 1970s and '80s came the first small telescopes in space that observed at different wavelengths of light—infrared, ultraviolet, X-ray, and gamma-ray. This phase of development culminated in the 1990 launch (and 1993 repair) of the Hubble Space Telescope.

THE OUTER LIMITS

What have all these telescopes shown us? At the start of the century, we thought we lived at the center of our galaxy, but by 1920 Harlow Shapley had moved us out to the edge. In 1929, Edwin Hubble showed us that we were part of a universe that was expanding. In 1930, Clyde Tombaugh enlarged our Solar System with his discovery of Pluto; then, after the Second World War, Walter Baade's work showed that the universe was virtually double the size that it had previously been thought to be.

The outer limits of the universe have been pushed back continuously since then, a major milestone occurring in 1963, when Maarten Schmidt discovered the first of many quasars—the intensely energetic centers of far-off galaxies. By the 1980s, astronomers were discovering the bubble-like distribution of clusters of galaxies in space.

A key to the origin of the universe—the microwave background radiation—was discovered in 1965 by Arno Penzias and Robert Wilson, while in 1992 the COBE satellite telescope discovered the faint imprint of the origin of the galaxies in this radiation.

1970-1974	1975-1979	1980-1984	1985-1989	1990-
1971–2 First detailed close-range pictures of Mars (Mariner 9).	**1975** First images from the surface of Venus (Venera 9)	**1980** First detailed study of Saturn and its system (Voyager 1).	**1985-6** Halley's comet returns, observed by the Giotto and Vega spacecraft.	**1990** Hubble Space Telescope launched and found to be defective.
1973 First fly-by of Jupiter (Pioneer 10).	**1976** Vikings 1 and 2 land on Mars.	Commissioning of Very Large Array Radio Telescope in New Mexico.	**1986** Voyager 2 fly-by of Uranus.	Magellan spacecraft begins radar mapping of Venus.
	1977 Discovery of minor planet Chiron in the outer Solar System by Kowal.		**1987** Supernova 1987A appears in the Large Magellanic Cloud.	**1992** COBE satellite discovers structure in microwave background.
1973 150 inch (4m) telescope commissioned at Kitt Peak, Arizona	Discovery of the rings of Uranus. 		**1989** Voyager 2 fly-by of Neptune.	Keck 400 inch (10 m) telescope commissioned.
1974 First close-range pictures of Mercury and Venus cloud tops (Mariner 10).	**1978** Discovery of Pluto's satellite, Charon, by Christy.	**1981** Second fly-by of Saturn (Voyager 2).		**1993** Hubble Space Telescope repaired in space.
	1979 Two fly-by missions to Jupiter (Voyagers 1 and 2). First fly-by of Saturn (Pioneer 11).	**1983** Infrared Astronomical Satellite (IRAS) completes the first full survey of the infrared sky.	Geller and Huchra announce evidence of walls and voids in the distribution of galaxies.	**1994** Periodic Comet Shoemaker-Levy 9 collides with Jupiter.

CHAPTER TWO
STARS *and* GALAXIES

The universe is studded with stars and galaxies of infinite variety. Understanding a little about them enhances our enjoyment of the beauty of the night sky.

THE EMPTINESS *of* SPACE

By looking at the following images and comparing their relative sizes,
we can step out into space and begin to conceive of
the sheer vastness of our universe.

LANDSAT VIEW OF WASHINGTON DC *The Potomac River can be seen here dividing Maryland from Virginia. The city areas appear blue, while vegetation has been colored pink. To convey an idea of the scale involved, the width of this image covers approximately 30 miles (50 km) in each direction.*

EARTH *as photographed by the European meteorological satellite, Meteosat. The view of Washington DC (above, left) would occupy less than 1/64 inch (0.3 mm) of this image of our 7,927 mile (12,760 km) diameter globe. The color in this picture has been added to simulate the Earth's natural appearance. The first people to be far enough away to see the whole Earth from space were the astronauts of the Apollo 8 mission, in 1968.*

EARTH AND MOON, *as viewed by the Galileo spacecraft on its way to Jupiter. On this scale, the Moon would be 60 inches (1.5 m) behind the Earth. Other planets have moons, but only the Earth and Pluto have moons that are a large fraction of the planet's size. Contrast and color have been enhanced, especially for the Moon, to improve visibility.*

THE SOLAR SYSTEM *The Sun and its family of planets are featured in this portrait (orbits and planetary sizes not to scale). The blue ring near the center is the Earth's orbit, 93 million miles (150 million km) from the Sun in the center. Pluto's angled orbit, crossing Neptune's, is about 40 times larger. Looking back from this perspective, our tiny Earth is all but lost from view.*

NEAREST STARS *(below) These stars, all lying within 20 light years of the Sun, make up our solar neighborhood. On the scale of this illustration, the stars would really be infinitesimal points the size of atoms.*

1 Proxima Centauri, Alpha Centauri A, Alpha Centauri B; 2 Barnard's Star; 3 Wolf 359; 4 Sirius A, Sirius B; 5 Epsilon Eridani; 6 Epsilon Indi; 7 61 Cygni A, 61 Cygni B; 8 Tau Ceti; 9 Procyon A, Procyon B

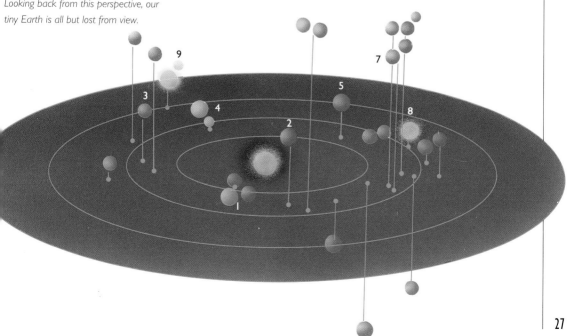

Put three grains of sand inside a vast cathedral, and the cathedral will be more closely packed with sand than space is with stars.

SIR JAMES JEANS (1877–1946), English astronomer

THE MILKY WAY GALAXY, *seen here as a glowing maelstrom of some 200 billion suns, is about 150,000 light years across. Our Sun is no more than a minute speck in one of the galaxy's great spiral arms. The solar neighborhood shown in the previous image would be a tiny patch just $1/1,000$ of an inch (0.03 mm) across. At the speed of light, it would take 28,000 years to get from the local neighborhood to the center of the galaxy.*

DISTANCE SCALES

While a mile or a kilometer is a suitable unit of distance here on Earth, when we say that the Andromeda Galaxy (M 31), the nearest large galaxy to the Milky Way, is 9 quintillion miles (21 quintillion km) away—that's 9,000,000,000,000,000,000 —the number is so large that it becomes meaningless.

ASTRONOMICAL UNITS

A useful measure in our Solar System is the Astronomical Unit (AU), the average distance between the Earth and Sun—approximately 93 million miles (150 million km). Mercury is $1/3$ of an AU from the Sun, while Pluto averages 40 AU from the Sun.

LIGHT YEARS

For measuring distances further afield, we use the light year—the distance light travels in a vacuum in a year. Traveling at about 186,000 miles (300,000 km) a second, light can travel the equivalent of seven times around the Earth in a single second. A light year is about 6 trillion miles (10 trillion km).

PARSECS

Astronomers also use a parsec, equivalent to about 3.3 light years or 206,000 AU, as a unit of distance. If an object were only one parsec away, it would have an annual parallax (see below) of one second of arc (about $1/1,800$ the diameter of the Moon).

PARALLAX

In astronomy, the term parallax refers to the apparent displacement of a nearby star against the background of more distant stars. This is not caused by the motion of the star itself, but by the Earth's motion as we observe the sky from different parts of our orbit around the Sun.

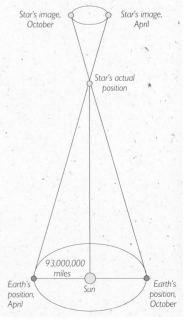

Star's image, October

Star's image, April

Star's actual position

93,000,000 miles

Earth's position, April

Sun

Earth's position, October

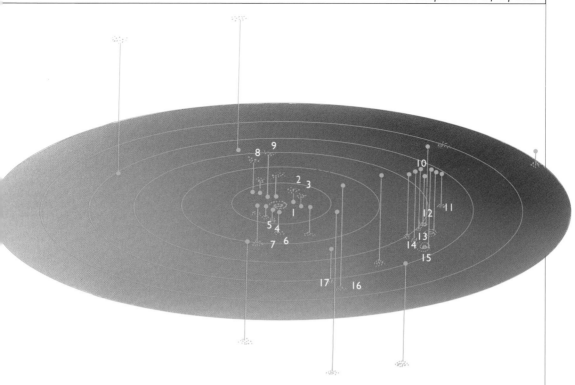

THE LOCAL GROUP *is a small cluster of some 30 galaxies within 2.5 million light years of the Milky Way. The Milky Way and Andromeda are the dominant galaxies within it. The galaxies are shown at about double correct size relative to their distance from one another.*

1 Milky Way; 2 Draco; 3 Ursa Minor; 4 Small Magellanic Cloud; 5 Large Magellanic Cloud; 6 Sculptor; 7 Fornax; 8 Leo I; 9 Leo II; 10 NGC 185; 11 NGC 147; 12 NGC 205; 13 M 32; 14 Andromeda (M 31); 15 M 33; 16 IC 1613; 17 NGC 6822

UNIVERSE OF GALAXIES *This view from the Hubble Space Telescope shows a cluster of galaxies of all types, one of thousands of such clusters scattered across the sky. From this far out, our entire Milky Way Galaxy is hardly more than a small blob of light.*

29

DWARFS, GIANTS, *and* SUPERGIANTS!

With the exception of the Moon and the planets, every fixed point of light in the sky is a star—a nuclear powerhouse—and these stars range from dwarfs to supergiants.

SIRIUS A *is a main sequence star some 10,000 times brighter than its white dwarf companion Sirius B, seen here as a dot 11 arc seconds away. Special methods to enhance the contrast produced this image.*

Our Sun is a star, and it appears large and bright only because we are close to it. Most stars are too far away to appear as more than points of light, even in the largest telescopes, but we still know quite a bit about them. We know, for example, that they differ in size, and that at least half of them consist of two or more stars locked in a gravitational embrace.

WHAT IS A STAR?

Stars are large balls of hydrogen and helium, with a sprinkling of other elements, all in gaseous form. As gravity pulls the star's material inward, the pressure of its hot gas drives it outward, resulting in an equilibrium. A star's energy source lies at its core, where millions of tonnes of hydrogen are fused together every second to form helium. Although this process has been going on in our Sun for almost 5 billion years, it has used only a few percent of its hydrogen supplies. It is in the prime of its life—its main sequence phase.

TYPES OF STARS

Main Sequence Stars Early this century, two astronomers tried to make sense of the wondrous variety of stars. Ejnar Hertzsprung, from Holland, and Henry Norris Russell, from the United States, came up with the Hertzsprung–Russell (HR) Diagram that plots the temperature at a star's surface

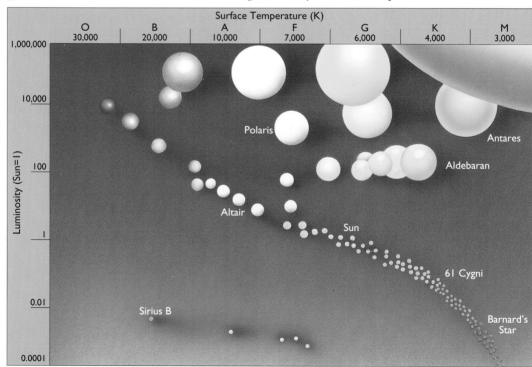

	O 30,000	B 20,000	A 10,000	Surface Temperature (K) F 7,000	G 6,000	K 4,000	M 3,000

Luminosity (Sun=1): 1,000,000 · 10,000 · 100 · 1 · 0.01 · 0.0001

Polaris, Antares, Aldebaran, Altair, Sun, 61 Cygni, Sirius B, Barnard's Star against its brightness, making allowance for its distance away from us.

Most stars, including the Sun, plotted nicely on a band across the diagram called the main sequence. These main sequence stars are often called dwarfs, although some are 20 times larger than the Sun and 20,000 times brighter.

Red Dwarfs At the cool, faint end of the main sequence are the red dwarfs, the most common stars of all. Smaller than the Sun, they are carefully doling out their fuel to extend their lives to tens of billions of years. If we could see all the red dwarfs, the sky would be thick with them, and the HR Diagram would be heavy with stars in its lower right corner. But red dwarfs are so faint that we can observe only the closest ones, like Proxima Centauri, the nearest star to Earth.

White Dwarfs Smaller than red dwarfs are the white dwarfs—typically the size of the Earth but the mass of the Sun. A volume of white dwarfs the size of this book might have a mass of about 10,000 tonnes! Their position on the HR Diagram marks them as quite different from their dwarf cousins. They are "stars" whose nuclear fires have gone out.

THE HERTZSPRUNG–RUSSELL DIAGRAM *(left) presents the basic groups of stars, together with an indication of their color and relative size. The relative numbers of stars in each portion of the diagram are not correctly presented. The majority of stars lie on the main sequence, stretching from upper left to lower right, with numbers increasing toward the faint, red end. Above the main sequence are many giants like Aldebaran and rare supergiants. Tiny, faint white dwarfs lie across the bottom.*

Red Giants After the main sequence stars, the most common stars are the red giants. They have the same surface temperature as red dwarfs, but are much larger and brighter and thus lie above the main sequence in the HR Diagram. These monsters typically have a mass similar to the Sun's, but, if they traded places with the Sun, their atmospheres would envelop the Solar System's inner planets. Most are, in fact, orange in color, but R Leporis, in Lepus, is so red that some have likened it to a drop of blood.

Supergiants Lying along the top of the HR Diagram are the largest stars of all—the rare supergiants. Betelgeuse, in Orion's shoulder, is close to 600 million miles (1,000 million km) across. Orion's other chief luminary is Rigel, a blue supergiant, which is one of the most luminous stars visible to the naked eye. Barely one-tenth the size of Betelgeuse, it is still almost 100 times the size of our Sun.

SIZE COMPARISON *The Sun, a dwarf, is portrayed here as a yellow globe. The red giant star behind it (of which we see only a small portion) is 100 times larger! By contrast, a white dwarf star would be less than ¹⁄₅₀ the size illustrated. A neutron star might be 1,000 times smaller again, represented here by the grossly oversized black dot.*

BETELGEUSE *This is the only star, other than the Sun, for which we are currently able to produce an image of the star's surface, using techniques borrowed from radio astronomy.*

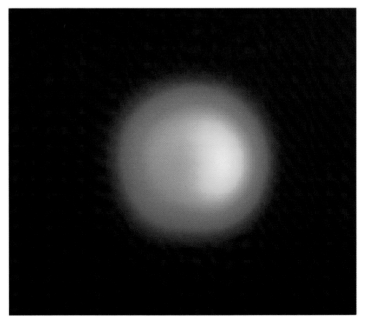

STARS *and their* EVOLUTION

Stars live a long time—way too long for us to observe a single star's life cycle. Instead, we build a picture of the life of a star by studying stars at different periods in their lives.

Look at the Milky Way on a dark night and you will notice bands of darkness scattered along it— vast clouds of gas and dust that only become visible when they block out the light from more distant stars. Spread throughout the galaxy, these giant clouds provide the raw material for new stars.

STAR NURSERIES
If you look through a pair of binoculars at Orion's sword, the middle star will appear misty. A telescope reveals a cloud of gas brightly lit by a group of bright blue stars—the latest in a series to be born from the gas permeating the sky in Orion. Their formation

was perhaps triggered by a star from an earlier generation exploding as a supernova within the last few million years.

Our Sun, like all other stars, is presumed to have been born in a similar association of clouds and stars, but its sibling stars have, over billions of years, dispersed in space.

ALPHA CENTAURI
After birth, stars live the majority of their lives as well-behaved main sequence dwarfs, like the three stars making up the nearby Alpha (α) Centauri system. The brightest star is much like the Sun and can expect to spend around 10 billion years on the main sequence, slowly growing warmer and brighter as it consumes the hydrogen fuel in its core. Its cooler, orange companion will do the same for perhaps 20 billion years, while the red dwarf Proxima Centauri may last for at least 60 billion years.

A STAR'S LATER LIFE
What happens when a main sequence star like our Sun runs low on hydrogen to fuse into helium? As the fuel situation becomes critical, the star becomes brighter, larger, and redder and begins to fuse helium into carbon. No longer a main sequence dwarf, the star is now a red giant.

The bright star Capella, in Auriga, has two components

THE LIFE OF THE SUN *Prompted perhaps by a nearby supernova explosion, a gas cloud collapses to form the Sun and the planets. After a long life on the main sequence, the Sun will pass through a red giant phase before gently ejecting its outer layers, ending its days as a faint white dwarf.*

about three times more massive than the Sun, both evolving into red giants. Several other bright stars in the sky have already become red giants, like Arcturus in Boötes and Aldebaran in Taurus.

MIRA THE WONDERFUL
In the constellation of Cetus, there is an old red giant called Mira that has roughly the same mass as our Sun. In the span of about a year this star becomes bright enough to be seen easily with the naked eye and then fades until you cannot see it without using a telescope. Several billion years from now, perhaps our Sun will regularly expand and contract in the same way Mira does.

Stars are like animals in the wild. We may see the young but never the actual birth, which is a veiled and secret event.

HEINZ R. PAGELS (1939–1988), American physicist

SUPERNOVA 1987A *photographed at the height of its outburst in early 1987. Overlaid in black are images made before the explosion, showing the outburst centered on a previously catalogued twelfth magnitude blue supergiant known as Sanduleak -69°202. Pre-explosion images of supernova stars are extremely rare.*

CELESTIAL GRAVEYARDS

A star's life depends largely on how massive it is,

as does its ultimate fate as a white dwarf, neutron star, or black hole.

A star like the Sun will live as a red giant for about a billion years. Then, as it exhausts its nuclear fuel, it will cast off its outer layers to form a planetary nebula (see p. 43) surrounding a blazing hot core. This core will slowly fade away over millennia as a cooling white dwarf. Something like this is the fate of most stars; however, very massive stars end their lives far more spectacularly in a supernova explosion.

SUPERNOVA 1987A

On February 24, 1987, Ian Shelton, a Canadian astronomer at the University of Toronto's telescope in Las Campanas, Chile, had just begun looking for novae and variable stars. His targets were the two galaxies closest to the Milky Way—the Large and Small Magellanic Clouds.

Shelton was only two days into his program when he noticed an extra star near the Tarantula Nebula in the Large Magellanic Cloud. It was so bright that he realized it must be a supernova—the explosion of a dying star. Since the last bright supernova had been seen in 1604, Shelton could hardly believe what he saw. Astronomers monitored Supernova 1987A over the next few months as it became brighter and brighter until it was almost as luminous as all the other stars in the Large Magellanic Cloud put together!

To explode as a supernova, a star probably starts out with at least 10 times the mass of the Sun. It evolves into a red

A PLANETARY NEBULA, *such as the Dumbbell Nebula (M 27) in the constellation of Vulpecula, represents the final moment of glory for most stars. The central core fades to obscurity as a white dwarf.*

supergiant, producing elements as heavy as iron in its nuclear furnace while also shedding some of its distended envelope. Unless it loses a large amount of its mass, the star eventually becomes unable to resist the relentless force of gravity. In a mere fraction of a second, the core collapses, blowing the star apart.

The expanding blanket of material which formed the bulk of the star collides with the surrounding interstellar medium to produce a supernova remnant. The Crab Nebula (M 1), in the constellation of Taurus, is all that is left of a mighty explosion that many people witnessed in 1054.

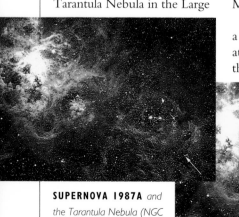

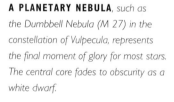

SUPERNOVA 1987A *and the Tarantula Nebula (NGC 2070), as seen in 1984 (above), and in 1987 (right)*

A BLACK HOLE (right) *is believed to emit energy from a surrounding accretion disk and ejected jets, as portrayed in this artist's impression.*

A PULSAR *in the Crab Nebula (M 1) in Taurus is seen here in X-rays as it pulses on and off 30 times every second.*

NEUTRON STARS AND PULSARS

After an explosion, all that remains is the collapsed core, known as a neutron star, with density far higher than even that of a white dwarf.

Neutron stars rotate very rapidly, emitting beams of light and radio waves which, if they sweep past the Earth, appear like light from a cosmic lighthouse. The resultant blinking has led astronomers to call these pulsars (from "pulsating stars"). The fastest pulsars rotate at almost one thousand times per second!

BLACK HOLES

There is yet another fate that can befall a massive star—it can become a black hole.

The blue supergiant HDE 226868, in the constellation of Cygnus, attracted the attention of astronomers when they discovered that its position coincided with a powerful X-ray source, Cygnus X-1, that flickered once every few thousandths of a second. It also appeared to orbit an invisible companion every 5.6 days. The mass of this unseen object was calculated to be between 8 and 16 times the mass of the Sun—too massive to be stable as a neutron star. Instead, it would collapse indefinitely until it finally disappeared altogether, leaving only a source of gravity so strong that even light could not escape from it. Most astronomers conclude that that is what had happened to the object alongside HDE 226868—it had collapsed to form a black hole. Gas from its companion star, HDE 226868, is raining down on the black hole to produce the X-rays.

ROBERT EVANS

Searching for supernovae was a province left almost entirely to professional astronomers until the Reverend Robert Evans began to search for them in the late 1950's from his home in Hazelbrook, Australia.

As he became familiar with galaxy after galaxy, by examining them whenever the weather was clear (on a single night he might check more than 200), he learned to spot intruding stars. In one particularly productive year he made more that 15,000 galaxy checks, ending the year with a bounty of two supernovae for all his work.

Finding supernovae is clearly not easy. Most galaxies have bright starlit regions that look like exploding stars but on closer examination they turn out to be regions of gas. It takes persistence to put so much effort into a visual search, a quality that Bob Evans has in good supply.

Behold, directly overhead, a certain strange star was suddenly seen . . . Amazed, and as if astonished and stupified, I stood still.

TYCHO BRAHE (1546–1601), Danish astronomer

STELLAR COMPANIONS

As a single star, our Sun is in a minority:

more than half the stars in the sky have at least one companion in space.

Stars usually seem to occur as pairs, triples, or even clusters. Lone stars like the Sun are in a minority, although it generally takes the magnification of a telescope to reveal a star's multiple nature.

Sometimes nature tries to fool us by lining up two stars in our sky which are in fact widely separated in space. These optical doubles can be distinguished from true binaries only by observations spread over many years which may eventually reveal the actual motion of the stars. The orbital motion of true binaries will also eventually be apparent, although this may take hundreds of years.

Most binary systems are so closely held together that we have no hope of discerning their two components visually. Instead, we use instruments such as the spectrograph, which closely examines the colors of the light received to read telltale binary signatures.

DOUBLE STARS

The nearest example of a double star is Alpha (α) Centauri. Visible best from southern latitudes, this magnificent pair is the closest stellar system to the Sun. A naked-eye view shows only one star—the third brightest star in the sky—but through a telescope you can easily see two. Strictly speaking, Alpha Centauri is a triple, since the system has a third member, Proxima Centauri—a small red dwarf that orbits the other two just over 2 degrees away.

Mizar, the middle star of the Big Dipper's handle, is one of the

OPTICAL VERSUS PHYSICAL DOUBLES *A chance optical alignment is far different to a pair physically in orbit about each other.*

MASS EXCHANGE *can occur in many close binaries. Here, a blue star leaks material slowly to the accretion disk around a black hole.*

best known double stars. Close by is a fainter star named Alcor that can be seen on a clear night. These two stars do not form a true binary system—in our line of sight, Alcor just appears to be 12 arc minutes from Mizar. However, if you point a telescope at Mizar itself, you should be able to see that it is two stars, forming a true double (with a separation of 14 arc seconds).

Observing double stars with a small telescope can be fun and quite challenging: they test the resolving power of your telescope as well as the steadiness of the seeing. A 2.4 inch (60 mm) refractor should be able to split stars of similar brightness separated by about 2 arc seconds in very good conditions at high magnific-

ALCOR AND MIZAR *(in the lower left corner) are perhaps the most famous pair of stars for northern observers. They are separated by over ⅓ of the Moon's diameter.*

RED GIANT LOSING MASS *to a white dwarf companion (below). The gain in mass may cause the white dwarf to erupt periodically.*

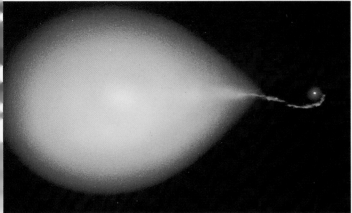

ation. A 4 inch (100 mm) reflector should do roughly twice as well—about 1 arc second. If you fail with a particular double, it may be that the atmospheric seeing could be a lot better, or your telescope is not performing up to par.

The coloring of the stars in a pair is always interesting, the contrast between them sometimes making the colors appear unusual. For example, Albireo, the star at the head of Cygnus, the Swan, is a splendid double star whose brighter member is yellowish, while the fainter component has a green tinge (separation 34 arc seconds). It can be fun to ask several people what colors they would ascribe to the stars of a given pair to see how their perceptions vary.

MULTIPLE STARS

Castor is the second brightest star in the constellation of Gemini, the Twins. Using a small telescope on a night with steady star images, you might

be able to "split" the star into its two components (separation 3.9 arc seconds). You may also be able to see a third, fainter member farther afield. By

analyzing the light from these stars as seen through a spectroscope, we know that all three stars are also double, so the system in fact consists of six members locked in a gravitational embrace.

The bright star Alpha (α) Capricorni is a double so wide (separation over 6 arc minutes) you may be able to "split" it without any optical aid. It is a chance alignment, but a telescope makes all the difference since both the major stars are themselves true doubles (separations 46 and 7 arc seconds), and the fainter of one of these is yet another double.

A SAMPLER OF EASY-TO-OBSERVE DOUBLE STARS

This list comprises some of the prettiest doubles in the sky, in addition to those already mentioned. Each is plotted on the Constellation charts in Chapter 5.

Beta (β) Cephei: Both white; magnitudes 3.3 and 7.9 (separation 1.3 arc seconds).

Gamma (γ) Andromedae: Through a moderate-sized telescope this can be seen as a beautiful pair, the brighter member orange, the fainter blueish (separation 10 arc seconds).

Theta (θ) Orionis: The famous trapezium in the center of the Great Nebula in Orion. A small telescope will reveal four bright stars; a larger telescope will show two fainter ones as well.

Gamma (γ) Leonis: A beautiful pair that revolve about each other about once every four centuries (current separation 5 arc seconds).

Gamma (γ) Virginis: This pair revolves around each other in about 180 years. As the two stars circle each other the distance between them, from our point of view, appears to be shrinking (current separation 3 arc seconds).

Epsilon (ε) Lyrae: The famous "double double" in Lyra. Not far from bright Vega, each of these stars is itself a double (each with separation 2.5 arc seconds).

Alpha (α) Crucis: The brightest star in the Southern Cross (separation 5 arc seconds).

STAR CLUSTERS

Stars are grouped not just in pairs or triplets,
but also in clusters ranging from tens to
hundreds of thousands of members.

NGC 3293 *in Carina lies 8,500 light years away, and is made up of several dozen relatively bright stars.*

At the lower end of the clustering scale lie the open or galactic clusters. They are so called because they lie relatively close to us in the disk of the galaxy. They generally have only tens or hundreds of members, so we can see their individual stars clearly. Most of these stars are younger than our Sun, some being among the youngest stars we can see.

The Pleiades (also known as the Seven Sisters or M 45), in the constellation of Taurus, is a famous open cluster. You can see six or seven stars with the naked eye, but through a telescope you can see many fainter stars. Not far from the Pleiades is Aldebaran, the bloodshot eye of Taurus, the Bull; and the V-shaped grouping of stars near Aldebaran is the Hyades cluster, one of the closest of the open clusters.

It is often difficult to separate foreground or background stars from true members of a cluster. If, like the Hyades, the cluster is near enough to us, careful study allows the cluster stars to be picked out because they are moving with a common speed and direction across the sky. Our Sun is actually moving through an association of stars sometimes called the Ursa Major moving cluster, which is comprised of most of the Big Dipper stars and some others scattered across our sky, as revealed by their common motion.

THE PLEIADES *(M 45) is probably the best known open cluster, plainly visible to the naked eye in Taurus. The surrounding dust is only revealed clearly in photographs such as this.*

GLOBULAR CLUSTERS

Scattered throughout the sky lie more than 100 globular star clusters. These gigantic aggregations of stars are perhaps 15 billion years old—as old as the Milky Way itself. Through a small telescope they look like small fuzzy balls, but larger instruments (with 8 inch [200 mm] apertures or more) resolve the balls into untold thousands of stars.

Globular clusters have been studied almost since the invention of the telescope: Abraham Ihle found the large cluster in Sagittarius we now call M 22 as early as 1665. In 1786 William Herschel noted that these clusters were mottled, "which denotes their being resolvable into stars".

Most of the globular clusters we see are of our galaxy, but not in it—they are in its halo, or outskirts. The Southern Hemisphere has the best of these distant giant hives of old stars, one of which, Omega (ω) Centauri—a large oval clustering of hundreds of thousands of stars—is visible to the naked eye and can be seen on spring evenings from the southernmost parts of the United States. 47 Tucanae, perhaps the finest globular cluster of all, reserves its glory solely for people in the Southern Hemisphere.

The best cluster in the Northern Hemisphere is M 13 in Hercules. Twenty-three thousand light years away, but about 100 light years across, this

cluster is easy to find on a side of the "keystone" in Hercules.

What would it be like to live inside a globular cluster? The sky would be awash with hundreds of stars as bright as Vega, and when night came the sky would be a twilight one. Perhaps the best part of the view from many of the

GLOBULAR CLUSTER 47 TUCANAE
(NGC 104) is one of the glories of the southern sky and shows a pronounced central core of stars.

globulars, though, would be seeing the glorious form of the Milky Way galaxy, its spiral arms spreading out to cover half the sky.

. . . the perceptible Universe

exists as a cluster of clusters,

irregularly disposed.

Eureka,
EDGAR ALLAN POE (1809–49),
American writer

SELECTED OPEN CLUSTERS

Name	Constellation	Magnitude (total)	Approx. size (arc min.)	Approx. distance (light yrs)
M 44 (Beehive)	Cancer	3.1	95	600
NGC 4755 (Jewel Box)	Crux	4.2	10	6,800
M 39	Cygnus	4.6	30	7,300
M 35	Gemini	5.5	30	2800
h (The Double Cluster)	Perseus	4.4	35	7,000
Chi (χ)		4.7	35	8,100
M 6 (Butterfly)	Scorpius	4.6	26	1,500
M 7	Scorpius	3.3	50	800
M 11 (Wild Duck)	Scutum	5.8	14	5,600
M 45 (Pleiades)	Taurus	1.2	110	400
(Hyades)	Taurus	0.5	330	150

SELECTED GLOBULAR CLUSTERS

Name	Constellation	Magnitude (total)	Approx. size (arc min.)	Approx. distance (light yrs)
M 3	Canes Venatici	6.0	18	35,000
NGC 5139 (Omega (ω) Centauri)	Centaurus	3.7	36	17,000
M 13	Hercules	5.9	17	23,000
M 92	Hercules	6.5	11	26,000
M 15	Pegasus	6.4	12	34,000
M 22	Sagittarius	6.0	18	10,000
M 5	Serpens	5.8	17	26,000
NGC 104 (47 Tucanae)	Tucana	4.5	31	16,000

VARIABLE STARS

Variable stars change in brightness.

Some pulsate in size, while others violently eject material.

Still others are pairs that just "get in each other's way".

O ver the years, amateur astronomers have followed the cycles of changing brightness of hundreds of distant variable suns. Undoubtedly one of the best known is Mira (Omicron [o] Ceti), whose variation was discovered in 1596 by David Fabricius, a Dutch pastor and skilled amateur astronomer. Mira is a red giant star, about the mass of the Sun, which varies in brightness over a period of 11 months. At its brightest, it is easily seen with the naked eye, then it grows progressively dimmer, dropping below naked-eye

ECLIPSING BINARIES *are closely spaced pairs, orbiting each other in hours or days. They periodically obscure parts of each other, as seen from Earth, causing a decline in the total light we observe. The smaller member of the pair is usually brighter, causing the stronger dip in light to occur when its surface is obscured. The exact way the light varies reveals that very close pairs pull each other out of shape.*

visibility, and you wonder if it will disappear forever. Eventually however, the decline halts, then the process reverses, and finally, one night, you can see it with the naked eye once more.

MIRA STARS

With several thousand examples, long-period variables like Mira are the most common variables known. They are red giants that pulsate, but over periods of hundreds of days, and they cycle far less regularly than other variables. Their range of

variation in apparent brightness is often as much as 6 or 8 magnitudes—a factor of several hundred—although the stars actually change in size by less than 50 percent.

ECLIPSING BINARIES

Some variables are double stars that are aligned in such a way that one passes in front of the other, and then behind it, resulting in the light from the

NOVA CYGNI 1975 *appeared in August 1975, peaking at magnitude 1.8. The nova is at the very center of this image taken four months later.*

SOME VARIABLE STARS TO LOOK FOR

Name	Type	Magnitude range	Period (in days)
Eta (η) Aquilae	Cepheid	3.5–4.4	7.2
R Carinae	Mira	3.9–10.5	308.7
R Centauri	Mira	5.3–11.8	546.2
Delta (δ) Cephei	Cepheid	3.5–4.4	5.4
Omicron (o) Ceti	Mira	3.4–9.3	332.0
Zeta (ζ) Geminorum	Cepheid	3.7–4.2	10.2
Delta (δ) Librae	Eclipsing binary	4.9–5.9	2.3
Beta (β) Lyrae	Beta Lyrae type	3.3–4.3	12.9
Beta (β) Persei	Eclipsing binary	2.1–3.4	2.9
R Scuti	RV Tauri type	4.5–8.2	140

system varying. The most famous of these is Algol in Perseus. Every 2.9 days, Algol's fainter member passes in front of the brighter one. During the resulting eclipse, lasting 10 hours, the system's brightness drops about 1 magnitude.

CEPHEIDS

Cepheid variables are named after Delta (δ) Cephei, the first to be discovered. Like a Mira star, a Cepheid's fluctuations result from a cycle of changes inside the star, causing it to expand and contract. As a Cepheid expands in size, over a day or more, it dims, and as it contracts it brightens. This takes place with clockwork regularity. The northern pole star, Polaris in Ursa Minor, is a Cepheid with a small range of fluctuation. Its variations seem to have been dying recently, indicative of the fact that variables are not permanently variable—they are simply in an unstable phase.

The Cepheids' major claim to fame is that they provided the key to the size of our galaxy. In 1784, a deaf-mute teenager named John Goodricke discovered Delta (δ) Cephei's changes. More than a century later, Harvard's Henrietta Leavitt studied the cycles of about 25 Cepheids in the Small Magellanic Cloud, one of the galaxies closest to us, and found that the brighter their average magnitudes were, the longer were their periods of variation. Then Harlow

Shapley made a great intuitive leap regarding all Cepheids: for any two Cepheids with the same period of variation, the one with the brighter average magnitude will be closer to us. This period–luminosity relationship, as he called it, became an elegant yardstick for measuring distances in space.

CATACLYSMIC VARIABLES AND NOVAE

While all variables put on performances, the cataclysmic or eruptive ones offer surprises. Once thought to be new stars (hence the name), novae are in fact explosive outbursts in binary star systems.

Novae consist of a large star and a small, hot star—usually a white dwarf—which pulls a stream of gas from the larger star. The captured gas grows hotter as more gas is transferred and then finally explodes in a thermonuclear detonation, leading to an increase in brightness of as much as 10 magnitudes. The stars are not disrupted, however, and ordinary novae are

believed to repeat this process in cycles of hundreds or thousands of years. Smaller eruptions are typical of "dwarf" novae such as SS Cygni, in Cygnus, which can brighten by as much as 4 magnitudes in a few hours, doing so at intervals of between 20 and 90 days.

R Coronae Borealis is a rare example of what appears to be a "reverse nova", although it is actually an unrelated phenomenon. Instead of brightening every so often, it normally stays at its maximum brightness—easily visible through binoculars—until, it is thought, an eruption of soot-like carbon smothers it for several weeks or months to the point where it cannot be seen even through a small telescope.

CEPHEID VARIABLES *change in brightness because they actually pulsate in size (exaggerated here), also changing slightly in surface temperature, and hence color. A typical "light curve" (in red) shows the star's brightness peaking quickly and then slowly declining over a period, typically, of a few days.*

HENRIETTA LEAVITT *(1868–1921) was an American expert in the photographic analysis of variable star brightness. She applied this to her discovery of the law relating period and brightness of Cepheid variable stars, which gives us clues to their relative distances from us.*

CLOUDS AMID *the* STARS: *the* NEBULAE

Of all the sights in the sky, the delicate clouds of gas and dust known as nebulae—the birthplaces and graveyards of the stars— are among the most stunning.

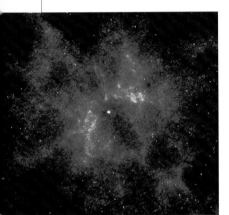

T he Latin word for cloud, nebula, is the term used for the cosmic gas and dust that lie among the stars. Most nebulae are lit up by light from stars within them. The biggest nebulae do not shine at all— they can only be seen when they block the light of more distant stars.

EMISSION NEBULAE

Emission nebulae are the most colorful nebulae of all, glowing like neon signs from energy released by the stars within them. Under a good sky, a large telescope (of 10 inch [250 mm] diameter or more), will reveal the glowing gas and enable you to see tinges of red and green in some nebulae,

THE ROSETTE NEBULA *(NGC 2237-9) in Monoceros is a large nebula surrounding the open cluster NGC 2244. It is spectacular in photographs but quite faint to the eye at the telescope.*

PLANETARY NEBULAE *are not always ring shaped. This Hubble Space Telescope image shows the faint eleventh magnitude NGC 2440 and its 16th magnitude central star.*

but only long-exposure photographs really unveil the range of colors—particularly the striking red characteristic of hydrogen gas.

The easiest nebula to see is the middle star of the sword in the constellation of Orion— the Great Nebula (M 42). If the sky is dark you can see it as a misty spot with the naked eye, but even from a city, binoculars will clearly reveal it as a fuzzy cloud. The larger the telescope, the more beautiful the view becomes.

REFLECTION NEBULAE

Reflection nebulae, as their name suggests, are lit by reflected light from nearby stars. Merope, one of the stars in the Pleiades, is surrounded by the classic blue glow of a reflection nebula. You cannot see it without a telescope, but on a clear, dark night out in the country a small telescope should reveal the faint cloudiness near Merope.

A DARK NEBULA *(below) dimming and blocking light from the distant curtain of stars of the Milky Way in one dark patch, with more dark dust scattered elsewhere in the image*

DARK NEBULAE

"There's a hole in the sky!" More than two centuries ago, William Herschel, discoverer of Uranus and one of the greatest observational astronomers, discovered a new kind of object. Herschel thought that these objects might be "old regions that had sustained greater ravages of time" than the surrounding stars. His son, John Herschel, thought that they looked like gateways to something beyond. Deep in the southern sky, right beside the Southern Cross, he studied a dark region that looks so black we now call it the Coal Sack.

The Herschels had stumbled on dark nebulae—gas and dust

with no nearby stars to light them, hiding the stars behind. They are usually seen against either the background of the Milky Way's stars or the brightly lit gas of a nebula.

PLANETARY NEBULAE

Many, but not all nebulae are the birth-places of stars. After a star like our Sun evolves into a red giant, it enters a brief phase in which it blows off its outer layers. Eventually these layers become visible as a thin shell of

THE TRIFID NEBULA *(M 20) in Sagittarius features red emission and blue reflection regions, plus the dark lanes which give the object its name.*

gas around it. Nineteenth-century astronomers noticed that some were the shape and color of the planets Uranus and Neptune, so they called them planetary nebulae. We now know that these nebulae have nothing to do with planets.

The Ring Nebula (M 57) in Lyra is the best known of these nebulae. Visible in small telescopes as an out-of-focus star, in a 3 inch (75 mm) or larger one it looks like a soft ring. The Clownface or Eskimo Nebula in Gemini, another famous planetary, really does look like the face of a clown in a larger telescope, complete with a large bright nose at the center.

THE CONE NEBULA *in Monoceros is a spectacular dark nebula silhouetted against bright emissions but, like some other famous nebulae, is difficult to see with small telescopes.*

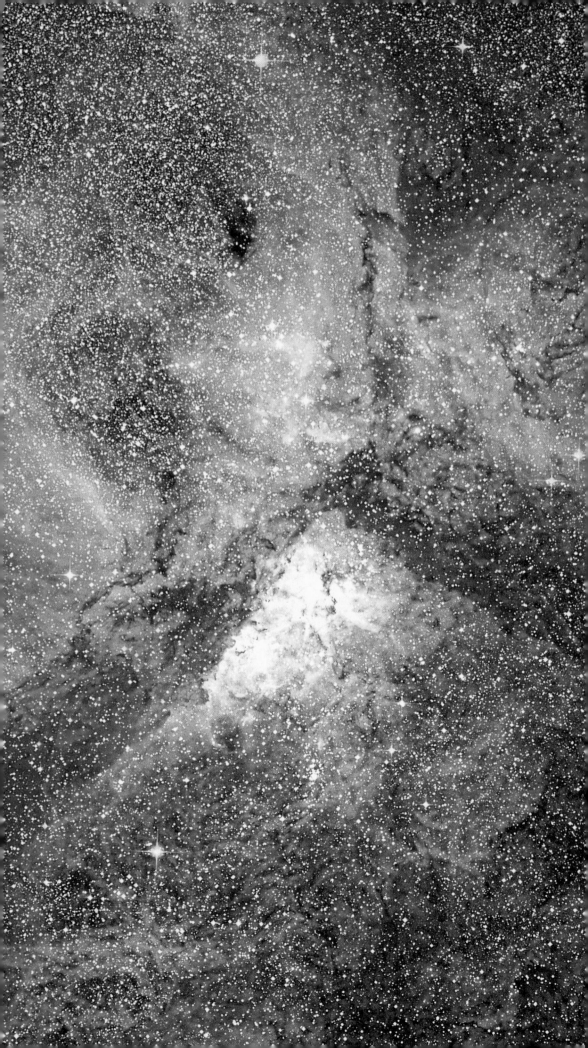

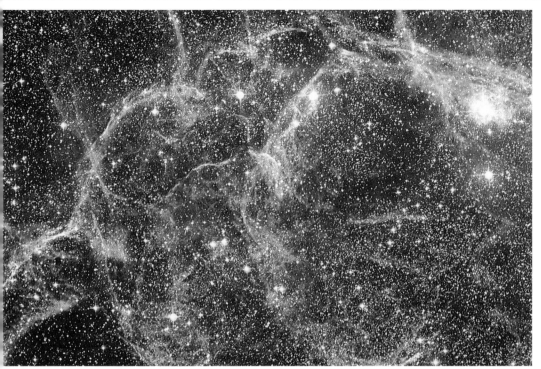

SOME NEBULAE TO LOOK FOR

Name	Constellation	Type	Mag. (total)	Approx. size (arc min.)	Approx. dist. (lt yrs)
NGC 3372 (Eta (η) Carinae)	Carina	Emission	6	120 × 120	3,700
NGC 2070 (Tarantula)	Dorado	Emission	5	40 × 25	180,000
M 57 (Ring)	Lyra	Planetary	8.8	2 × 1	5,000
NGC 2237 (Rosette)	Monoceros	Emission	6	80 × 60	3,000
M 42 (Great Nebula)	Orion	Emission	3.5	85 × 60	1,500
M 8 (Lagoon)	Sagittarius	Emission	5	80 × 35	4,500
M 17 (Omega)	Sagittarius	Emission	6.9	45 × 35	5,500
M 20 (Trifid)	Sagittarius	Emission, reflection, and dark	7	29 × 27	3,500
M 1 (Crab)	Taurus	Supernova remnant	8.5	8 × 6	4,000
M 27 (Dumbbell)	Vulpecula	Planetary	7.3	8 × 4	3,500

SUPERNOVA REMNANTS

A considerably more massive star than the Sun dies much more violently—in a supernova explosion—and the gas ejected from the star sweeps up other gas from the interstellar medium to form a supernova remnant. The most famous of these is the Crab Nebula, in the constellation of Taurus, which looks like an oval glow in smaller telescopes. It exploded almost 1,000 years ago, and photos taken over a period of a few years show that it is still expanding.

THE VELA SUPERNOVA REMNANT *(above) is what remains of an explosion that occurred 10,000 years ago. It is likely that the supernova that produced these luminous shards of gas was bright enough to have been seen in the daytime.*

THE ETA CARINAE NEBULA *(left) (NGC 3372) is one of the glories of the southern sky. It is an enormous star-forming region 20 times the size of the Orion Nebula, and appears about 2 degrees across in our sky.*

BOK GLOBULES *(right) are dense clumps of dust, seen against background emission, which may eventually collapse to form stars.*

THE MILKY WAY

Almost all the stars, star clusters, and nebulae visible in a small telescope are part of our local system of stars—the Milky Way Galaxy.

THE ORIGIN OF THE MILKY WAY
by Tintoretto (1518–94) shows Juno breastfeeding the infant Hercules. According to this legend, droplets of milk spilt upwards and became the stars of our Milky Way Galaxy (below).

For many thousands of years, the Earth was thought to be at the center of everything, until Copernicus and Galileo persuaded us that the Sun ruled the heavens. By the eighteenth century, astronomers were wondering whether the Sun itself was near the center of the system of stars known as the Galaxy or the Milky Way (galaxias is Greek for "milky"). And was this the only galaxy?

In 1917 Harlow Shapley, at Mount Wilson in California, used the distances to variable stars in the far-off globular clusters to show that the Sun is, in fact, some 50,000 light years away from the center (now revised to about 30,000 light years). In 1924, Edwin Hubble had shown that the Milky Way Galaxy was just one of many.

SPIRAL STRUCTURE

Since then, scientists have been struggling to describe what the Milky Way looks like from the outside. This is a little like trying to paint a picture of the exterior of a house when all you have seen is the inside of a room or two. But studying nearby buildings from your window may give you a clue as to how your house might look.

After Shapley established that the Earth is far from the center of the Milky Way, astronomers wondered if our galaxy were pinwheel-shaped like many of our neighboring galaxies—M 31 in Andromeda and M 33 in Triangulum, for instance. We now know that the Milky Way is indeed a spiral galaxy that is flat, except at the center where there is a wide bulge. It contains some 200 billion suns, many of which we never see because so much dust and gas obscures our view.

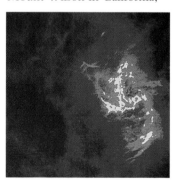

THE GALACTIC CORE *is seen here using radio waves. The spiral-like structure is a glowing cloud of gas some 10 light years across, with a bright point of radio emission at its heart.*

The disk of the Milky Way is about 1,500 light years thick, with spiral arms uncoiling to a distance of about 150,000 light years. Each star and nebula in this vast array orbits the galaxy's center more or less independently, our Sun completing an orbit in about 240 million years. Surrounding this galactic disk is a halo of older stars that stretches perhaps another 150,000 light years.

THE SPIRAL STRUCTURE *of the Milky Way Galaxy as it would appear from the outside—an artist's impression, of course!*

WHAT IS AT THE CENTER?

Because the Milky Way consists of so much gas and dust, it hides its secrets well, one of the greatest being what lies at its center. For some time astronomers thought that a source of strong radio emissions called Sagittarius A was at the center. Now an even smaller source of intense radiation, known as Sagittarius A★, has been found in this complex region. It might be a vast black hole with the mass of millions of suns. Material cascading into it would release the huge amounts of energy we can detect.

BART BOK AND THE BIGGER AND BETTER MILKY WAY

At the time that Harlow Shapley was redefining our place in the Milky Way, a Dutch youth named Bart Bok was longing to follow in his footsteps. By 1930, Bok had crossed the Atlantic to Harvard and was studying the Eta (η) Carinae Nebula in the constellation of Carina—one of the richest areas in the southern Milky Way.

In 1941, Bart and his wife Priscilla, also a well-known astronomer, published the first edition of their book *The Milky Way*, which has since become a classic. They continued their studies both of the Eta (η) Carinae region and the galaxy's spiral structure at Australia's Mount Stromlo Observatory and later at the University of Arizona. Happy to address any audience, Bok loved to talk about what he called "the bigger and better Milky Way"— a galaxy that becomes larger and more interesting the more we learn about it.

. . . Torrent of light and river of the air, Along whose bed the glimmering stars are seen, Like gold and silver sands in some ravine. . .

HENRY WADSWORTH LONGFELLOW (1807–82), American poet

47

GALAXIES

Beyond the confines of the Milky Way, space is filled with a myriad more galaxies, many in clusters with hundreds or even thousands of members.

The various types of galaxies are classified according to a scheme originated by Edwin Hubble in the 1920s.

TYPES OF GALAXIES

Spiral galaxies, such as the Milky Way, feature blueish spiral arms traced out by bright, young stars surrounding a nuclear bulge. Barred spiral galaxies are similar but feature a bar across the nuclear region, with the spiral arms trailing off the ends of the bar.

Elliptical galaxies show no spiral structure at all and range from flattened cigar shapes to spheres. Finally, there are the irregular galaxies, which do not seem to be highly structured; and the peculiar galaxies which are hard to classify, probably because they have suffered some sort of disturbance.

Although spirals appear to be the most common galaxies—mainly because great spirals like the Milky Way are relatively easy to identify—the majority of galaxies are actually low brightness irregulars and dwarf ellipticals.

Spiral Galaxies Closest of the great spirals, the Andromeda Galaxy (M 31) is the most distant object that we can see by unaided eye from Earth. Because we are viewing it from a sharp angle it looks like an elongated, misty spot rather than a spiral. However, the

SOME GALAXIES TO LOOK FOR

Name	Constellation	Type	Mag. (total)	Approx. size (arc min.)	Approx. dist. (million light yrs)
M 31	Andromeda	Spiral	3.4	180 × 60	2.3
NGC 5128	Centaurus	Elliptical	7.0	18 × 15	13
Large Magellanic Cloud	Dorado	Irregular	—	600	0.17
NGC 253	Sculptor	Spiral	7.1	25 × 7	10
M 33	Triangulum	Spiral	5.5	60 × 40	2.3
Small Magellanic Cloud	Tucana	Irregular	—	250	0.20
M 81	Ursa Major	Spiral	7.9	18 × 10	7
M 87	Virgo	Elliptical	8.6	7 × 7	40
M 104	Virgo	Spiral	8.3	9 × 4	40

SMALL MAGELLANIC CLOUD *(top left) is a nearby example of a small irregular galaxy, so close that it spans some 4 degrees in the southern constellation of Tucana.*

SPIRAL GALAXY *(above) M 100 is a fine example from the Virgo cluster of a large spiral galaxy, with spiral arms traced out by hot blue-white stars and glowing red gas, surrounding a yellower core of older stars. The arms lie within a fainter disk which, in the Milky Way, would include stars like the Sun.*

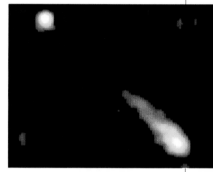

ELLIPTICAL GALAXY *M 87 in Virgo is a giant of its type, surrounded by globular star clusters which appear here as fuzzy "stars".*

QUASAR *(below) At around 2,000 million light years away, 3C273 is the closest of these luminous powerhouses. This radio image reveals a jet of material (lower right) emerging from the core of the quasar (upper left).*

Whirlpool Galaxy (M 51) in Canes Venatici, or M 81 in the constellation Ursa Major, are true pinwheels in shape.

Barred spirals are far less numerous than conventional spirals and it is unclear what makes a spiral produce a bar. M 83 is a bright galaxy in Hydra often classified as barred. NGC 1300 in Eridanus is rather fainter, but presents a classic barred structure.

Elliptical galaxies vary in size from dwarfs to giants. The giants are the largest galaxies we see, often dominating the center of a cluster, and probably growing from collisions with spiral galaxies. They pass through an intermediate stage showing evidence of collision, as seen in NGC 5128 in Centaurus, and finally settle down again as a giant elliptical, like M 87 in Virgo.

Irregular galaxies are mostly faint, amorphous groups of stars, much smaller than the spiral galaxies. The Small Magellanic Cloud, a satellite galaxy of the Milky Way, is a classic example deep in the southern sky.

Peculiar Galaxies Any galaxy that seems to have suffered some kind of severe disturbance is called a peculiar galaxy.

For example, the bright galaxy M 82 in Ursa Major is a maelstrom of bright starry regions covered by irregular lanes of dust. Its appearance probably results from a mammoth burst of star formation.

Seyfert galaxies and quasars Seyfert galaxies, of which M 77 in Cetus is the best known, are great spiral systems with bright centers. Named after Carl Seyfert, who first noticed them in 1942, these galaxies are part of a continuum of "active" galaxies with unusual, often violent, core activity.

A quasar or QSO (quasi-stellar object) is believed to be the highly energetic core of an active galaxy. They are thousands of times brighter than the rest of the galaxies they inhabit and thus are detectable over enormous distances. If we placed a large ordinary spiral galaxy, such as M 31 in the constellation Andromeda, at the distance of the quasars, it would be invisible to us.

EDWIN HUBBLE

In 1924, with the recently completed 100 inch reflector at Mt Wilson, Edwin Hubble (1889–1953) and Milton Humason were the first to resolve individual stars within the Andromeda galaxy (M 31), then show that it was a galaxy—one of countless many—like our own Milky Way. (See pp. 50–51.) In 1925 Hubble set up a system of classifying galaxies, which we still use in modified form today. He is also remembered for the law bearing his name, which gives us a measurement of the time scale of the universe and its ongoing expansion. His outstanding contributions to astronomy are recognized in the naming of the Hubble Space Telescope, launched by NASA in 1989.

THE BIG PICTURE

We live on the third planet out from a medium-sized sun,

about two-thirds of the way along one spiral arm of the Milky Way Galaxy.

Where does this put us in the universe as a whole?

Around the turn of the century, Vesto M. Slipher was studying the sky from the Lowell Observatory in Flagstaff, Arizona. The director, Percival Lowell, was interested in investigating planets around other suns, and he thought that the spiral "nebulae" that were being discovered could be stars with new planetary systems forming around them.

To test this theory, Lowell asked Slipher to study the composition of the spiral nebulae using a spectrograph, which splits the light up into its component colors. Using a 24 inch (600 mm) refractor, Slipher had to spend as much as two full nights gathering enough light for a spectrum of a single nebula. The results baffled him: all the spectra showed large shifts of the features toward red light.

It was the work of Edwin Hubble, at Mount Wilson in California, that eventually resolved the mystery of these "redshifts". With the great 100 inch (2.5 m) reflector at Mount Wilson at their disposal, Edwin Hubble and Milton Humason took such clear photographs of nearby spiral nebulae that by 1924 they were able to resolve them into individual stars.

In 1929, Hubble demonstrated that redshifts were telling us that the galaxies are moving away from us at hundreds or thousands of miles per second. At these speeds,

THE LOCAL GROUP *is a confederation of some 30 or so galaxies, dominated by two giant spirals—the Milky Way and the Andromeda Galaxy (M 31). Most members are faint dwarf ellipticals and irregulars that would be invisible from a greater distance.*

The Universe is an infinite sphere, the centre of which is everywhere, the circumference nowhere.

Pensées, BLAISE PASCAL (1623–62), French mathematician and natural philosopher

THE EINSTEIN CROSS, *as seen by the Hubble Space Telescope (below), is actually four images of one distant quasar produced by a galaxy (in the center) 20 times closer acting as a gravitational lens (as illustrated at left).*

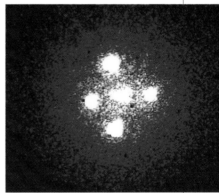

the light waves they leave behind are stretched out, noticeably reddening the light.

Hubble noted from his measurements that the fainter, and therefore, presumably, more distant galaxies, showed greater redshift. Thus, Hubble's Law states that the redshift of distant galaxies increases in proportion to their distance from us. Measuring redshift enables us to estimate distance in the universe.

THE DISTRIBUTION OF THE GALAXIES

Shortly after Hubble proposed that the universe was expanding, he claimed that the galaxies were distributed evenly in space. To prove his point, he took a large number of pictures of small regions of the sky using the huge Mount Wilson reflector. With the exception of what he called a zone of avoidance around the Milky Way, where dust obscured the galaxies behind it, he found galaxies in roughly equal numbers everywhere.

Other cosmologists disagreed with Hubble's findings. In a survey of the Northern Hemisphere sky taking wider fields of view, Harlow Shapley and Adelaide Ames noted considerable discrepancies in the populations of galaxies throughout the sky. In some places they found large numbers and in others relatively few. Clyde

Tombaugh, who discovered Pluto in 1930, confirmed Shapley and Ames's observations and went a step further by discovering, in 1937, a cluster of hundreds of galaxies in Andromeda and Perseus.

The greatest advance, however, came during the making of the Palomar Observatory Sky Survey with a new 48 inch (1.2 m) Schmidt telescope. Using its fine photographic plates, George Abell demonstrated that the galaxies are arranged unevenly into clusters and superclusters.

THE LOCAL GROUP

The Milky Way and the Andromeda Galaxy (M 31) are the largest members of a small group of up to 30 galaxies called the Local Group. This cluster is part of a supercluster of galaxies whose other members can be seen in the constellations of Coma Berenices and Virgo.

We now recognize other superclusters scattered across the cosmos, but are there clusters of superclusters? From recent observations with large telescopes, this seems unlikely. The superclusters appear to form enormous frothy structures in space, with vast "voids" between them. This immense expanding structure is partaking in the general expansion of the universe. Galaxies within clusters are gravitationally bound together,

but outside the clusters the expansion of space itself is inexorably pulling the clusters away from each other.

GRAVITATIONAL LENSES

At the end of the 1970s, a pair of identical quasars were found on a Palomar Sky Survey photograph, with a faint but massive galaxy lying between them. The galaxy and the quasar demonstrate part of Einstein's general theory of relativity—that gravitational sources can bend light. The gravity of the galaxy was acting as a lens, bending the light of the distant quasar so that it appeared double.

More exotic cases have since been discovered where galaxies are positioned in such a way that they cause more distant objects to break up into beautiful arcs and even rings, and one is so perfectly placed that the quasar far behind it is broken up into four images we call the Einstein Cross.

V.M. SLIPHER
(1875–1969) is known for his observations of the planets, nebulae, and distant galaxies.

CHAPTER THREE
SKYWATCHING TOOLS
and TECHNIQUES

*The sky offers a spectacle that everyone can delight in,
but with some practical tips and basic instruments,
your viewing can become an even richer experience.*

NAKED EYE ASTRONOMY

For tens of thousands of years we have enjoyed the spectacle
of the night sky without the benefit of binoculars or a telescope,
simply by looking upward in the dark.

RURAL OBSERVERS *These naked-eye astronomers were painted by Donato Creti (1671–1749).*

Without any optical aid at all, you can observe a wide range of phenomena in the sky. You can see the large dark areas on the Moon that are now known to have resulted from large objects crashing into it almost 4 billion years ago. So easy are they to make out that they have given rise to legends about there being a man, or a hare, in the Moon.

You can follow the nightly wanderings of five of the planets, and from one place on Earth or another you can spot the 88 constellations—the traditional star patterns. You can watch certain stars—the variables—alter in brightness over the course of days, weeks, or months. You can see star clusters like the Pleiades in Taurus and gas clouds like the Great Nebula in Orion. If the night is dark, the vast expanse of our own galaxy, the Milky Way, will be visible, winding across the sky, as will three of its neighboring galaxies.

There are also artificial satellites to be seen, and showers of tiny meteors. If you are far enough north or south, you can enjoy the fabulous fireworks display of the northern or southern lights (the aurora borealis and the aurora australis).

THE CHANGING SKY

The sky is constantly changing. The Moon rises about 40 minutes later, on average, every night, so as well as its phase differing each time it rises, its position relative to the stars also differs.

Although they move much more slowly than the Moon, the planets trace elegant paths among the stars and you can easily follow the movements of the five that are closest to us over a season of observing.

The stars themselves rise about four minutes earlier each night, which may not sound like much, but the difference adds up to about an hour every two weeks and a day over the course of a full year.

SKYWATCHING WITH CHILDREN
(right) is especially rewarding. You don't need a telescope to take a look at the wonders of the universe.

You can even see the sky alter during the course of a single night. In only two hours, stars that were near the eastern horizon will have risen to prominence high in the eastern sky, while others will have set in the west.

GOING OUTSIDE

For most of us, dusk marks the end of a day: for skywatchers (unless they watch the Sun), it marks the beginning. Although any clear night provides an invitation to go outside and see what is up, some nights offer special events, such as a lunar eclipse, when the Moon passes through the Earth's shadow. Meteor showers (which, like eclipses, are also forecast) provide another reason for turning off the TV and turning on to the sky.

A NAKED-EYE SPECTACLE *(right) Venus (top) and Jupiter (center) accompany the Moon. The eagle-eyed observer would also spy Mercury within the tree, just above the band of clouds.*

FIRST NIGHT OUT

When skywatching for the first time, it is best not to rush outdoors to begin learning the constellations as they are plotted on the Starfinder charts.

For a beginner, especially a child, the sky is a great spur to the imagination. On your first night out, try connecting the stars to make your own patterns and figures. Let your mind roam. It should be fun, and you are unlikely to forget the positions of the stars in your personal constellations. By doing this, you will start skywatching in an entertaining and memorable way.

CITY *and* SUBURBAN SKIES

Do you need a dark country sky for skywatching?
Not necessarily. Even in the city you can locate constellations,
observe the planets, and enjoy double stars.

BAD LIGHTING *This is a classic example—a sign illuminated from below, directing light up into the sky.*

Compared with a dark country sky, does the illuminated sky over a city or the suburbs have much to offer? Is looking up even worthwhile?

You bet it is. In some respects a bright sky is better for a beginner than a dark one, because you do not have to contend with a confusing undertow of 3,000 faint stars. It is easier to find the outlines of the major constellations when there are just a few dozen of the brightest stars in evidence, and observing the Moon and the bright planets is just as easy in the city as it is in the country. A group called the San Francisco Sidewalk Astronomers are well known for giving the public a chance to see what can be viewed through a telescope from within the city.

For a beginner, the best place from which to make your observations would be the familiar surroundings of your own backyard, porch, or even rooftop.

Any open space will do, but it is obviously best to be away from direct lighting, if possible. Choose somewhere that is close to home, then it will be easy for you to just take a few minutes whenever you feel like it to do a little skywatching.

LIGHT POLLUTION

Astronomer David Crawford, from the Kitt Peak National Observatory, has calculated that the United States spends about 2 billion dollars a year on lighting the underbellies of birds and airplanes!

The problem with cities, from an astronomer's point of view, is not so much the amount of lighting there is, but the direction in which it is pointed.

We need efficient lighting for road safety and security, but those who argue for this are in fact in agreement with those who long to have a darker sky. A brightly lit street can also be poorly lit, especially if the lighting is irregular, with alternating bright and dark areas. The answer lies in having well designed fixtures that direct the light downward and shield the bulb from view.

Some cities, such as Tucson, Arizona, have switched to using sodium lighting. These lights— especially the low pressure variety —emit single colors that astronomers can filter out, and they spread a diffuse light that is free from glare.

CITY LIGHTS *Sights such as this are not uncommon in the city. Both the buildings and the sky are equally brightly lit .*

CITY STARS *Orion's brilliant stars can be easily seen, even above the Vancouver skyline, although the photograph shows more stars than the eye alone might see.*

THE INTERNATIONAL DARK SKY ASSOCIATION

In 1987, David Crawford and Tim Hunter, an amateur astronomer from Tucson, Arizona, founded the International Dark Sky Association. This organization aims to foster an approach to city lighting that promotes safety and economy, as well as preserving the beauty of the night sky.

If you would like to persuade your local authorities to adopt better lighting practices, a group of you will need to do quite a bit of lobbying, but the International Dark Sky Association can provide assistance.

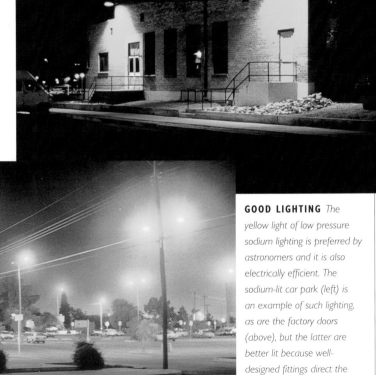

GOOD LIGHTING *The yellow light of low pressure sodium lighting is preferred by astronomers and it is also electrically efficient. The sodium-lit car park (left) is an example of such lighting, as are the factory doors (above), but the latter are better lit because well-designed fittings direct the light downward, which is where it is needed.*

OBSERVING TECHNIQUES

If you learn some simple measuring methods
and are able to take the time to prepare yourself,
you'll enjoy your skywatching to the full.

When we enter a darkened room after being somewhere well lit, we have trouble seeing until our eyes have become used to the reduced light. Called dark adaptation, this process takes at least 15 or 20 minutes, but it is time well spent if you want to prepare yourself to make a thorough observation. During this time the pupils of your eyes will gradually open to their fullest extent (as much as ¼ inch [7 mm] for children; probably only ⅕ inch [5 mm] for older people), to allow in more light from the stars.

Clyde Tombaugh was a farmer in Kansas when he became interested in sky-watching. To get the most out of the Kansas sky he used to first sit in a completely dark room for an hour, simply allowing his eyes to become adapted to the dark. Taking his hobby seriously paid off, for at the age of 22 he was

HOW MANY STARS? *If you adapt your eyes to the dark, it will enable you to see many more stars.*

taken on to work at the Lowell Observatory, and a year later he discovered the planet Pluto.

When you are adapting your eyes to the dark, avoid accidentally looking at bright lights. If you are in your backyard, turn off the porch lights and close the curtains so as to hide the light indoors.

Always protect your eyes from excessive light at other times. If you have been out in the bright Sun all day without sunglasses, your eyes will take much longer to dark adapt.

RED LIGHTS

Many amateur astronomers use dim red lights or red-filtered flashlights to keep their eyes dark adapted during observing

THE RED VIEW *A normal flashlight with a red plastic filter is an essential accessory when using the Starfinder charts in Chapter 5.*

sessions, as any bright white light will close the eyes' pupils again within seconds.

A piece of red plastic or cellophane fixed across the front of the flashlight is all you will need to do the job. Have the red light just bright enough to read by, but no brighter.

Whereas other animals hang

their heads and look at the

ground, he made man stand

erect, bidding him look up to

heaven, and lift his head to

the stars.

Metamorphosis, OVID,
(43 BC–AD 17?), Roman poet

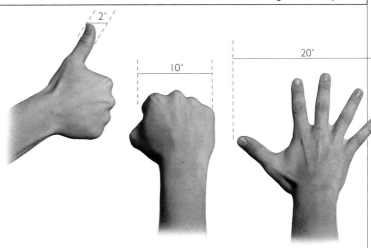

MEASURING SIZES AND DISTANCES

Astronomers use degrees, minutes, and seconds to measure sizes and distances in the sky. For example, 90 degrees is the distance from the horizon to the zenith point directly overhead. The Moon and Sun present disks of about ½ degree, or 30 arc minutes, in size. The finest detail your eye can resolve without an optical aid is about 1 arc minute, or 60 arc seconds.

An outstretched hand, held at arm's length, will be about 20 degrees wide from the tip of the thumb to the tip of the little finger—roughly the distance between the first and last stars of the Big Dipper (the

JUDGING DISTANCES *in the sky, using your hand at arm's length*

Plough) in the constellation of Ursa Major, which is visible from the Northern Hemisphere. Smaller distances can be measured with your fist at arm's length (about 10 degrees) and your thumb at arm's length (about 2 degrees). Your fist at arm's length would roughly cover the distance between the bright belt stars and brilliant blue–white Rigel in Orion, and your thumb would appear four times wider than the Moon.

After experimenting with this technique for a while, you will develop a feel for the distances between the various

"HANDS ON THE SKY" *Your hand can help you measure distances in the sky.*

stars. Later on you can use this approach to determine the relative sizes of the different constellations and to find your way around much larger regions of sky. The constellation charts in Chapter 5 have symbols indicating size in terms of the outstretched hand.

READY FOR ACTION *A hood connected to the jacket would be a worthwhile addition to this winter observing outfit.*

DRESSING FOR SUCCESS

For maximum skywatching enjoyment, it is wise to dress appropriately. Even in the summer you can become quite chilled if you are motionless for a long while outdoors, so always make sure you take a jacket or a sweater. And those ultra-frigid winter nights can be almost comfortable if you wear a snowsuit.

Protecting your head is most important. If you wear an insulated hat of some kind,

especially a hood connected to the rest of your outfit, you are bound to stay reasonably warm.

If you are planning to be outside for some time, take a groundsheet and a rug to sit or lie on, and maybe a sandwich and a vacuum flask of hot soup or coffee. A bag of nuts and dried fruit might also be welcome. In warm weather, the bugs are likely to be out, so you will need insect repellant.

Finally, be safety conscious. If you are on home territory, make sure you know where the obstacles are in your immediate surroundings. Its surprisingly easy

to collide in the dark with garden furniture or fall into the fishpond in the back yard. If you decide to drive some distance from home, let someone know where you are planning to go, what roads you intend to use, and when you expect to return.

CHOOSING *and* USING BINOCULARS

Binoculars are the most versatile viewing instruments.

We can use them to watch birds, a baseball game,

an opera, and—not least—the sky.

eye pieces

focussing knob

objective lenses

Binoculars are essentially two low-powered telescopes joined together, so that you can look through them with both eyes instead of just one. For certain aspects of skywatching they are the best instrument of all to use.

For the amount of money you would pay for a poor quality small telescope, you can purchase a well-made pair of binoculars with quality optics that will last a lifetime. Through binoculars you will be able to see the craters on the Moon, the moons of Jupiter, and five or ten times as many stars as can be seen with the naked eye.

One of the best ways to use binoculars is to be seated on a reclining garden chair, preferably an adjustable one,

PORRO-PRISM BINOCULARS
This is the style that is familar to most people. The shape of the binoculars results from the prism arrangement that produces an upright image that is the correct way round. Roof prism binoculars, on the other hand, are typically smaller and use a different arrangement of prisms to achieve the same result.*

so that you can set it high for views near the horizon, and low for observations higher in the sky. If the chair has arm-rests, you can rest your elbows on them to hold the binoculars steady, enabling you to see the stars as steady points of light rather than as a bunch of flitting fireflies. Another way of balancing your binoculars is to simply lean against a wall or fence.

Observers often mount their binoculars on a camera tripod, and you will certainly find a tripod necessary if you

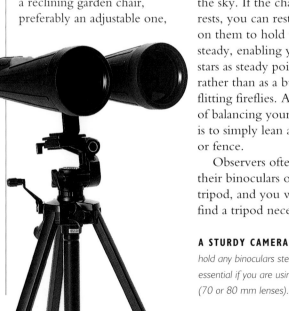

A STURDY CAMERA TRIPOD *will hold any binoculars steady, and is essential if you are using a larger pair (70 or 80 mm lenses).*

are using a larger pair, as they can be quite heavy and a strain to hold for any length of time.

TYPES OF BINOCULARS

Prisms are used in binoculars so that the image you see is around the right way. Other-wise it would be upside down and switched left to right, as it is in a simple telescope.

The two main styles of binoculars differ mainly in the orientation of the prisms that are used. The porro prism is simpler than the roof prism, and is the one most commonly used in binoculars today. The roof prism is more expensive, but is quite a bit lighter and is more compact.

PETER COLLINS

Arizonan Peter Collins searches the expanse of the Milky Way for the stellar outbursts known as novae, and has based his observing career simply on what he has been able to accomplish with a pair of binoculars.

In March 1992 he discovered his fourth nova—a faint star in the constellation of Cygnus that later became bright enough to be seen with the naked eye.

Like most astronomers who are successful nova searchers, Collins has somehow managed to memorize the patterns of the stars in the Milky Way, where almost all the novae appear.

He has accomplished this seemingly impossible task in an imaginative and engaging way by creating his own mini-constellations, each pattern of stars about 2 degrees across. One, for example, looks like one of William Herschel's extraordinary telescopes.

For Collins, binoculars are enough. They provide him with all he needs in order to enjoy the sky, know its star patterns intimately, and to watch the emergence of exploding suns. "The sky has a personality," he stresses.

TEST BEFORE YOU BUY

Many of the cheaper binoculars available from department stores are useless, for unless the two optical paths are exactly aligned, you will see slightly different images with each eye. As your eyes struggle to compensate, viewing soon becomes uncomfortable.

To test the collimation (alignment) of a pair of binoculars, ask someone to cover one of the objective lenses with a book or something similar while you look at a distant object. Both your eyes should be open. Then ask your friend to quickly remove the book. If you see two images that your brain then merges into one, then the binoculars are out of alignment.

It is best to buy your binoculars from a reputable dealer of optical equipment, who can probably also fix the collimation if a hard knock puts them out of alignment sometime later.

BINOCULARS *give great views and are easy to use, which is especially important for kids.*

WHAT SIZE BINOCULARS?

The performance of binoculars depends on the diameter of the objective lenses (the lenses at the front) as well as on the magnification that is provided by the eyepieces. Various combinations of lens and eye-pieces provide a considerable range of possibilities.

For night viewing, most experienced observers recommend using a pair of 7 x 50 binoculars, which always use the porro-prism design. The 7 refers to the magnification and the 50 refers to the diameter of each of the objective lenses (in millimeters).

The larger the diameter of the lenses, the more light they can collect and direct toward the eye, and thus the more effective they are in enabling you to pick out faint objects. While 10 x 50 or 8 x 40 binoculars are also suitable for using in the dark, they do not match the light that is available to the dark-adapted eye quite so well.

Binoculars with objective lenses up to 80 mm in diameter are commonly available, but a pair that large is quite heavy.

NEVER LOOK AT THE SUN THROUGH BINOCULARS *Never use your binoculars to observe the Sun: permanent blindness can result from even the briefest of looks. (Ways in which to view the Sun safely are described on pp. 64–5.)*

CHOOSING *and* USING *a* TELESCOPE

Telescopes have been of immense importance to astronomers since Galileo first turned his telescope on the Moon and Jupiter almost four centuries ago.

A 90 MM REFRACTING TELESCOPE
on a good quality altazimuth mount

lens cap

objective lens

finder

eyepiece

focussing knobs

flexible slow motion controls

tripod

Size and stability should be your main considerations when buying a telescope. As a rule of thumb, it is best to buy the telescope with the largest objective lens or mirror you can afford. Usually a reflecting telescope will provide larger optics than a refractor of the same size, since mirrors are cheaper to make than lenses. The result is more light coming into your eye and brighter images of faint stars and galaxies. Also make sure that the telescope is well mounted: a wobbly stand will make your observing sessions difficult and ineffectual.

Other factors may also influence your choice. For example, refractors tend to give sharper, brighter images than reflectors of the same aperture. This is because their optical elements are far less likely to slip out of alignment. However, refractors of about 4 inch (100 mm) aperture are physically far bigger than their reflecting cousins—an

important consideration if you want to move your telescope around outside.

If you have a limited budget, you might initially opt for a reflector with a 4 inch (100 mm) diameter mirror, whereas a more serious (or wealthy) skywatcher might choose one of 8 inch (200 mm) diameter or more. Buying a good quality 2.4 or 3 inch (60 or 75 mm) refractor might also be a wise choice for a beginner. These small refractors are sometimes considered too small to be useful, but many people start at this level, especially children.

Buy your telescope from a reputable dealer rather than a big department store. You will be given more assistance in making your choice, you can compare a wide range of

A REFRACTING TELESCOPE *has a convex objective lens at one end that gathers light, and an eyepiece at the other that magnifies the image formed by the lens.*

This is the type of telescope that is shown sticking out through the observatory dome in cartoons and comic strips.

THE REFLECTING TELESCOPE *was invented by Isaac Newton in 1671. Instead of having an objective lens at the top of the tube, a concave mirror at the bottom gathers light. This primary mirror sends the light back up the tube, where, in the Newtonian design, a small, flat secondary mirror intercepts it and sends it to an eyepiece at the side.*

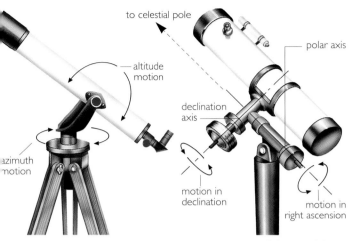

to celestial pole

altitude motion

azimuth motion

polar axis

declination axis

motion in declination

motion in right ascension

TYPES OF MOUNTINGS

The simple altazimuth mount is widely used on small refractors (far left) and is also found on many of the larger amateur reflectors. The equatorial mount is found on the majority of commercial reflectors, since it permits motorized tracking of the stars. Equatorial mounts come in a number of forms, including the German-style (left) and the fork mount that is commonly used on Schmidt-Cassegrain telescopes.

equipment, and you can be more confident about the quality of what you buy.

TYPES OF MOUNTINGS

A telescope without a good mounting is like a car without tires; it just will not work. The mounting is there to help you point the telescope, move it smoothly, and keep it steady. Most telescopes use either altazimuth or equatorial mounts. True to its name, the altazimuth pivots on two axes. One axis provides motion up and down (the altitude movement) and one turns the telescope around (the azimuth movement that parallels the horizon). As the object you are watching moves out of the telescope's field of

view, because of the Earth's rotation, you simply move the telescope about both axes to keep up with it.

With the more sophisticated equatorial mounting, you can follow a star with just a single motion as the Earth's rotation carries it across the sky. The mount is set up so that its polar axis points toward the north or south celestial pole—the points in the sky about which the stars

appear to rotate. This aligns the axis about which the telescope moves exactly parallel with the Earth's rotational axis. Adding a motor drive to the mount to move the telescope at a constant speed (15 degrees per hour) will counteract the Earth's motion and enable you to keep most of the objects you wish to look at in the field of vision for long periods of time.

CATADIOPTRIC TELESCOPES are reflecting telescopes that feature a correcting lens at the top to help form the image. Most of the catadioptric telescopes on sale are of the Schmidt-Cassegrain design. After the light has passed through the corrector, it reflects off the primary mirror and then off the curved secondary mirror, before finally passing through a hole in the main mirror and reaching the eyepiece. Although these telescopes are quite expensive, they are popular because they are small and easy to carry.

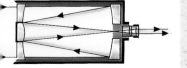

ISAAC NEWTON

Isaac Newton graduated in mathematics from Cambridge University in 1665, the year the Great Plague struck. Forced to return home because the university had been closed, Newton carried out all sorts of original research, and by the time he returned to Cambridge two years later he had developed laws that described how planets orbit the Sun, and a theory on the nature of light, color, and rainbows.

In 1671, he built a new kind of telescope that used a mirror to reflect the light it gathered, instead of a lens to refract it.

Despite his prodigious accomplishments, it took two

decades, and the support of his friend Edmond Halley, before he refined and published his work on gravitation.

SETTING UP YOUR TELESCOPE

I After you bring your telescope home, check that all the parts have been included. Set it up indoors and become familiar with its knobs and controls. Make sure the tripod and the telescope are firmly attached to the mounting.

2 Line up the finder with the telescope. This is best done during the day by viewing a distant treetop, or at night by looking at a street light. The adjustment can be done on a star, but it will be more difficult because of the star's motion across the sky.

3 If your telescope has an altazimuth mount you can simply take it outside, put it down in any position and start observing. If it has an

equatorial mounting, the polar axis of the mounting must be pointed toward the north or south celestial pole (see illustration below).

This is easily done in the Northern Hemisphere by finding Polaris, in the constellation of Ursa Minor. Sigma (σ) Octantis, the 5th magnitude southern pole star, is much harder to find. To make it easier to find the pole, make sure that the mounting is level; that the polar axis points upward from the horizontal at the angle of latitude of your site; and that the polar axis points north–south (true north and not magnetic north).

This may sound complicated, but it is often only necessary to get the alignment approximately right and you will be able to follow the stars easily, for many minutes at least.

4 Insert the eyepiece with the longest focal length. This will give you the lowest magnification and the widest field of view, and thus the best chance of finding things.

5 Choose a bright object like the Moon, a bright planet, or

a star. Center the star in the finder, and then look through the main telescope. If the finder is not perfectly aligned with the main telescope you will need to pan the telescope very slowly—up, down, and across. This might take time, but eventually you will find the star (and be convinced that aligning the finder carefully is worth the trouble). Once this happens, lock the mounting's axes. This miraculous moment is called first light—the moment the telescope first sees light from an object in the sky.

6 The star will probably look like a blob of light to begin with, or maybe like a doughnut if you have a reflecting telescope, which will mean the eyepiece needs focussing. Move the focussing knob one way; if the blob or the doughnut grows bigger, move it the other way until the star focusses into a point. Success!

N̲ever look through your telescope directly at the Sun without proper protection. Even a split second of unfiltered light could permanently blind you. And make sure you also warn family and friends about the dangers.

If your telescope has been sold to you with an eyepiece filter, throw it away. These filters should never be used.

Never use the finder to find the Sun either, since looking at the Sun through an unfiltered finder is also extremely dangerous. It is best to simply cover the finder up.

To observe the Sun safely, this is what you should do. Look at the way the shadow of the telescope is falling on the ground and move the telescope around

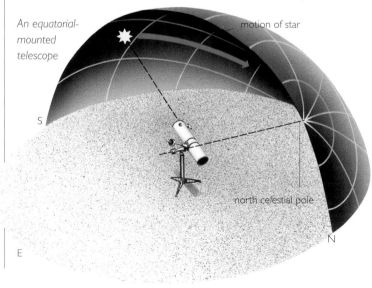

An equatorial-mounted telescope

motion of star

S

north celestial pole

N

E

THE MYTH OF HIGH POWER

In skywatching, high magnification is not always the thing you should be aiming for. Generally the highest power most observers would use is 60x per inch (2.4x per mm) of telescope mirror diameter, so 240x would be about the maximum for a 4 inch (100 mm) diameter telescope.

The higher the power, the harder it is to find what you are looking for and to keep it in the field of view. Moreover, high powers are far more sensitive to distortions in the image resulting from turbulence in the Earth's atmosphere. This effect is called seeing.

Even on a very clear night, severe twinkling of the stars is a sign of poor seeing. High powers would be useless on such a night, since the image would bounce around and be blurred by the unsteady air, making it impossible for you to detect any details.

AVERTED VISION

Appreciating the sky's fainter gems takes a certain amount of practice. The first time you look at Saturn, for instance, you may barely notice its rings, but as you become more experienced you will notice the dark zone where the rings are separated from the planet, and perhaps you will then be able to make out the gap between the two main rings.

OBSERVING THE MOON *through an equatorial-mounted refractor*

These subtle characteristics are easier to spot if you look "through the corner of your eye", a technique known as averted vision. To do this, shift your center of attention slightly, while concentrating on the object. Fainter details will become visible since you will be using the more sensitive rods in the retina of your eye, rather than the cones at its center.

VIEWING THE SUN SAFELY

until its shadow is at a minimum. The Sun should then be shining down through the telescope and out through the eyepiece, toward the ground. Project this image onto a piece of paper. Shield this "screen" from direct sunlight by fixing another sheet of paper around the eyepiece so that the Sun's image through the telescope can be seen clearly. Focus the eyepiece until the image is sharp.

An additional advantage of using this method is that it enables a group of people to look at the Sun easily and safely.

SAFE VIEWING *This telescope is equipped with a Sun projection screen. The Sun's image is focussed onto the lower screen, while the upper screen shields the image from other sunlight.*

Whenever you use this viewing technique with children, an adult should keep a close watch on the telescope to make sure that no one risks an eyeful of Sun. Cover the telescope aperture when it is not in use to prevent heat build-up in the telescope and any possibility of damage.

An alternative to the projection method is to use a solar filter well secured to the front of your telescope. The best ones are made from reflective mylar or glass. People may suggest you use filters such as exposed film or welder's glasses, but these are best avoided because of the possibility of an accident. Your eyes are far too precious to put at risk.

FINDING THINGS

For the first few weeks with your telescope, concentrate on the objects that you can see easily, particularly the Moon and the bright planets. By far the easiest thing to find, the Moon rewards you with a different show each night. You can travel along the crater rims, climb the lunar mountains, and explore the valleys.

As your experience grows, you are likely to want to start looking at fainter objects—ones that you will have to consult a Starfinder chart in order to find (see Chapter 5). Converting dots on a page into stars in the sky is a process that takes some getting used to, so do it slowly. First find a small group of bright stars on the chart, then find them in the sky. Proceed star by star, from the

IMPROVING MAGNIFICATION?
Observing Halley's Comet in 1909!

map to the sky, until you find the spot that contains the object you are looking for.

Using the telescope's finder, center on a star that is close to that object—preferably within 1 degree (or a fingertip held at arm's length). The finder usually has a field of view of about 6 degrees. Now use the main telescope and the lowest power eyepiece to move the telescope slowly in the direction of the object you are seeking. Finding it might take a few tries, but if it is a fingertip away from the star with the naked eye, it should be no more than one or two fields of view away in the telescope.

ACCESSORIES

Eyepieces

Eyepieces are used with a telescope to provide magnification. A wide variety is available, giving an extensive range of magnifications and fields of view.

Quite often telescopes are sold with only one inexpensive eyepiece and you will have to buy additional ones. They are not sold according to power, since the magnification depends on the telescope to which they are attached, but are available in a variety of focal lengths, measured in millimeters.

The lower the focal length, the higher the magnification.

Most beginners will find a 25 mm eyepiece most useful, plus perhaps a 12 mm one. On a typical 2.4 inch (60 mm) refractor, these eyepieces would yield magnifications of 28x (shorthand for 28 times) and 60x respectively. The same eyepieces on a typical 4 inch (100 mm) reflector would yield 40x and 85x. The full Moon would not quite fit in the field of view at 85x.

A word to the wise: don't pay attention to claims that a small telescope will offer powers of several hundred times. A limit of around 100x is more realistic. A good small reflector at only 75x will show you some fine views of the sky.

Finally, a few points about the various types of eyepieces. For example, 25 mm eyepieces are available in several sizes (and prices). The main differences are in the quality of the optics and the field of view. Eyepieces with a bigger barrel diameter are generally superior in both respects and will produce a better image, although they are more expensive and will not fit in smaller telescopes.

FIELDS OF VIEW *To the naked eye (right), the ½ degree disk of the Moon is surprisingly small, less than the width of your finger held at arm's length. Through binoculars (center) more detail is apparent, though the magnification is usually only around 7x or 10x. Only with a telescope (left), with magnification of 50x or more, does the Moon fill the field and reveal its heavily cratered surface.*

ACCESSORIES
Eyepieces (back) and colored eyepiece
filters. The Barlow lens (lower left) is used
with an eyepiece to double its
magnification. The camera adaptor
(right) couples a camera to a telescope.

Solar filters

Many small telescopes are sold with solar filters that fit in or near the eyepiece. Never use one of these as it can cause severe eye damage. If you have one, throw it out!

If you want to use a solar filter, buy a large one that can be attached to the front of the telescope. They are safer. Projecting the Sun's image, as described on pp. 64–5, is another alternative.

Finder or Telrad

Most astronomical telescopes come with a finder attached to them. The French name for this instrument, le chercheur, elegantly describes its function, which is to search through a wide area of sky to center the telescope on an object. The finder provides a wide field of view (about 5 degrees) at only 5x to 10x magnification.

The Telrad, a recently developed alternative, appears to project a faint red bulls-eye toward the sky, which you then center on the object you are searching for.

Slow-motion controls

Slow-motion controls enable you to move the telescope slowly and smoothly about either of the axes on its mounting. These controls are immensely helpful in making the fine adjustments necessary for finding an object and for keeping it centered in the telescope's field of view.

With equatorial mounts, simply turning a slow-motion control on the polar axis will keep your object in the field of view as the Earth's rotation works to move it away.

Motor drives

A motor drive enables your telescope to follow a star as it moves across the sky. If you have a drive on an equatorial mounting, you can watch an object for several minutes without having to touch the telescope at all, because the drive turns the slow-motion control on the polar axis for you. However, for the drive to be useful, you must have the telescope's polar axis pointed pretty accurately to the north or south celestial pole.

Computer control

Most of the more expensive telescopes can now be controlled through a computer. With such a system you simply key in a series of instructions and the telescope will then find whatever you wish (see p. 70).

I think that searching for a particular object is half the fun of skywatching, and if you leave all the work to your computer you lose the opportunity to learn about the sky for yourself.

MAINTENANCE AND CLEANING

Providing you treat all your optical components carefully, keeping them covered whenever possible, they will seldom need to be cleaned.

Generally a little dust or a small mark on the lens of a refractor will barely affect the image. Try blowing the dust off first, perhaps using a camera "blower brush". If the mark persists, try cleaning the lens with a damp, lint-free cloth in a gentle circular motion, as you would a camera lens. Stubborn marks will require some pure ethanol wiped on with a lens tissue and then wiped off with a new lens tissue. The trick is not to leave a smear. Never take the lens apart to clean it.

Cleaning the mirror in a reflector telescope is a major operation that should be undertaken rarely. First remove the mirror from the telescope, following the instructions in the manual. Remember that this mirror has the aluminum coating on the front, which means that you can easily scratch it. Move the mirror around in a bowl of warm soapy water, rinse it in clear water and remove any drops by touching, not rubbing, the surface with a small, lint-free cloth. Then, reinstall the mirror and align it so that it points light correctly to the eyepiece.

Eyepieces are much more likely to need cleaning than the telescope's lens or mirror, since they tend to gather dust from being dropped and fingerprints from hands groping around in the dark. These can be cleaned in the same way you would clean a refractor lens.

ASTROPHOTOGRAPHY

*Capturing the sky on film is rather more complicated
than taking a snapshot—patience and know-how are the key.*

Photographic film is so sensitive that in ordinary daylight an exposure of a hundredth of a second or less is enough to capture sufficient light to form an image. When photographing the night sky, however, you need exposures lasting from several seconds to as much as an hour.

All you need in order to start is some basic equipment: a camera with a setting (often called B or T) to allow long exposures; a shutter release cable, and some suitable film.

STAR TRAILS

Photographing the sky is complicated by the fact that the Earth rotates toward the east, resulting in everything in the sky moving westward. A conventional exposure will therefore show stars as curved lines, unless you can counter the Earth's rotation.

Why not start simply and photograph stars trails? Your results will be particularly dramatic if you use color film, as they will appear as short colored lines. Try a fast day-light print or slide film with an ISO speed of about 200.

Choose a daylight scene for your first shot. This will help the darkroom people cut the whole film correctly. Otherwise, unless you ask that the processed film not be cut at all, you might find they slice your faint pictures down the middle!

Mount the camera securely on a tripod. Attach a cable shutter release to the camera, set the focus on infinity, the shutter speed to B or T, and the aperture at its widest setting (perhaps f/1.8). Open the shutter with the cable release. Try a 30 second exposure and then release the shutter. Advance the film and then expose it for five and ten minutes. If you are using a standard 50 mm focal length lens, the first shot will show no trails, but longer exposures will show progressively longer ones.

STAR TRAILS *These two photographs of the stars around the "pointers" to the Southern Cross were taken from the same place—the mouth of a cave in South Australia.*

A short (50 second) exposure (above) shows the stars in their usual form—as points of light. A long (2 hour) exposure produces luminous star trails (right).

THE PLANETS

When photographing the planets, a telescope eyepiece in the camera adapter will relay the image from the telescope to the film and provide the necessary magnification. The exposure required varies according to the brightness of the target, the magnification, and the type of film being used. Slow film, such as ISO 50, is best for recording the fine detail of the small planetary image. Proper focussing is essential and it may be easier to focus on the Moon or a star than the planet itself.

CLUSTERS AND GALAXIES

Obtaining pictures of star clusters and galaxies is a great challenge. An exposure of at least 15 minutes will usually be needed and it is worth trying a variety of film speeds.

GUIDED PHOTOGRAPHY

Guided photographs take practice and more equipment. You need to attach a camera to your telescope, either on top, to photograph wide patches of the sky, or in place of the eyepiece, for magnified views of stars and planets. The latter requires a special adapter and a camera from which the lens can be removed. You also need a motorized equatorial mount. Finally, even the best motor drives will not guide perfectly, so you will need to monitor the pointing of the telescope using either a guidescope (a small telescope mounted on the main telescope) or an "off-axis" guider which allows you to see a small portion of the camera's view. A drive corrector can then be used to keep the star perfectly centered.

For long exposure astrophotography, the mounting must be aligned more precisely than for visual observing. This will save you major headaches later. Some companies manufacture special alignment telescopes that can be attached to your mount's polar axis.

DAVID MALIN

At the Anglo-Australian Observatory in Australia, the signature of astronomer David Malin is on some of the finest photographs of celestial objects ever taken. In astrophotography, as in any other kind of photography, choosing a subject and composing it is only half the task. Miracles in the darkroom are also needed, and Malin is a master of these.

He is a pioneer of a technique known as unsharp masking, in which a slightly out-of-focus positive film copy of a negative is exactly aligned with the negative. In the resulting print, the picture's fainter parts stand out, without the brighter sections being overexposed. Malin's beautiful work, featured throughout this book, brings out the most subtle differences in color and shape.

PERSONAL COMPUTERS and the STARS

Where once skywatchers found objects for themselves in the sky, now a computer can do it all for them.

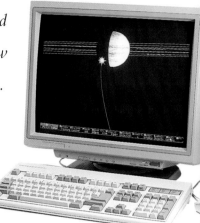

oday we can attach a computer directly to our telescope. Once we have keyed in the instructions, the computer will move the telescope to wherever we choose, beep when it's reached the right spot, and then flash a screenful of information telling us everything we are likely to want to know about what we are looking at.

STAR MAPS

Of the many computer programs designed for skywatchers, some of the most popular are essentially star charts combined with a catalogue. The best ones provide a map of any portion of the sky and a range of accompanying information. If, for example, you select the constellation of Orion and then click the cursor on top of Rigel, details may be given about Rigel's brightness, how far away it is, when it rises on that particular night, and so on. The program will also show you Rigel in the sky at any time of the day or night, from any place on Earth, for hundreds of years into the past or future.

COMET CRASH *A computer monitor displays the impact of Comet Shoemaker-Levy on Jupiter's night side as seen from nearby*

MAPS OF THE SOLAR SYSTEM

Other programs (often even the same program) will plot the positions of all the planets from several viewpoints. For your observing site on Earth, you are provided with a plot of the night sky with the positions of the Moon and the planets. Alternatively, you can choose to leave the Earth and the plane of the solar system altogether to see how the planets, comets, and asteroids are arranged in their orbits round the Sun.

IMAGING PROGRAMS

The availability of NASA spacecraft data, in particular, has made a vast array of images available to anyone with a computer and an image display program. For those with access to computer networks such as CompuServe or the Internet, vast archives of images and astronomical programs are available worldwide.

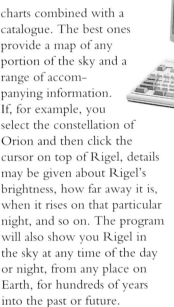

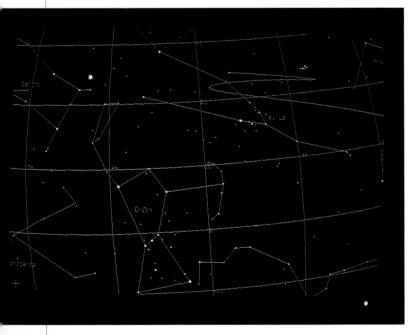

STAR CHARTS ON A SCREEN *This map of Orion and the surrounding constellations shows only the brighter stars. Note the constellation patterns, the coordinate grid, and the looping path of a planet (upper right) in Taurus.*

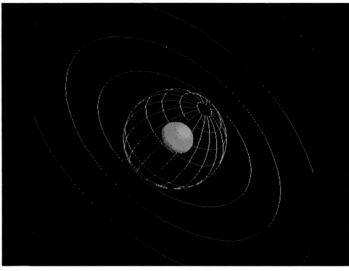

URANUS CLOSE UP *Some programs give you close-up detail of planetary spheres. Uranus, shown here, is featured with a coordinate grid, which helps you pick out the poles and watch them enjoy their 42-year-long "day" (see p. 252).*

CHARGE-COUPLED DEVICE CAMERAS

A charge-coupled device (CCD) is the light-sensitive electronic device that lies at the heart of modern video cameras. Some CCDs are particularly well suited to recording images of faint astronomical objects. They are so effective, in fact, that they can record the image of a nebula or galaxy as much as 30 times faster than photographic film.

Given that it is now possible to attach a simple CCD camera to your telescope for less than $1,000, clearly a revolution in amateur astronomy is underway.

HOW CCD CAMERAS WORK

When you open the shutter on a CCD camera, light falls on an electronic chip divided into tiny square picture elements that are known as pixels. Each pixel collects the amount of light it needs for its part of the picture. After the exposure has been completed, the chip transfers its information to a computer and the monitor then displays the result.

Since pixels vary in their ability to gather light, you need to "flat field" your image. This is done by having the computer compare the image, pixel by pixel, with an image you took earlier of a blank field—perhaps a twilight sky. CCDs also have a certain level of background signal. This can be reduced by cooling and almost eliminated by subtracting, in the computer, a "dark field" image taken with all light blocked from the CCD.

SEEING THE IMAGE

Details hidden in a CCD image can be revealed by using an imaging program to alter the brightness or contrast of the picture. For example, if the image shows details at the center of a galaxy, the fainter arms may be invisible. If you then adjust the display, it will brighten the image so that you can see the gas clouds of the arms, but the galaxy center will be too bright to distinguish clearly. Sophisticated programs can manipulate the image so that you can see detail in both the galaxy's center and its arms.

M 51 ON FILM AND CCD *The Whirlpool Galaxy (M 51), in Canes Venatici, captured on film (left) and shown by a CCD imaging program (above). The CCD image shows more detail in the bright core of the galaxy and in its companion. The computer program lets you zoom in on features of the galaxy and adjust the brightness and contrast in ways that are difficult to achieve in the photographic darkroom.*

CONTRIBUTIONS *of* the AMATEUR

A small number of amateur astronomers practice

their hobby with diligence and expertise

and produce professional results.

Most of the discoveries in astronomy today are made by professional astronomers who work with medium to large telescopes. Nevertheless, amateurs play a valuable part in modern astronomy because they provide the continuous monitoring and broad sky coverage that professionals cannot afford.

Amateurs routinely monitor thousands of variable stars and they often detect outbursts of the more irregular variables. Many comets, novae, and supernovae have been discovered by amateurs who dedicate an enormous amount of time and experience to careful, systematic searches of the sky.

THE CENTRAL BUREAU FOR ASTRONOMICAL TELEGRAMS

If you discover something new during your skywatching, such as a nova or a comet, the Central Bureau for Astronomical Telegrams (CBAT) will want to hear about it. Producing about 300 postcard-sized circulars a year, the CBAT announces discoveries by amateurs and professionals, publishes the predicted positions of comets and asteroids, and discusses observations made of everything from meteors to stars that emit X-rays.

Amateur astronomers are urged, however, to verify their findings before reporting them, as more than 90 percent of the reports of new comets

AMATEUR TELESCOPE *An 8 inch (200 mm) Schmidt-Cassegrain with the Moon, Venus, Saturn, and Jupiter.*

have proved to be false alarms. Ask an experienced amateur to repeat your observation or repeat it yourself the following night, to see if any alteration or movement has continued.

LESLIE C. PELTIER AND VARIABLE STARS

On a May evening in 1915, Leslie Peltier, an Ohio farm boy, was walking outdoors. "Something," he wrote, "perhaps a meteor—caused me to look up for a moment. Then, literally out of that clear sky, I suddenly asked myself, 'Why do I not know a single one of those stars?'" Deciding to save money to buy a telescope, he picked 900 quarts of strawberries on the family farm, at 2 cents a quart, to raise the necessary $18. Three years later, thrilled by their antics, he started observing variable stars, an activity he was to pursue for the rest of his life.

Peltier's reputation as a skilled observer spread, and he was offered the use of a 6 inch (150 mm) refracting telescope by Henry Norris Russell of the Princeton Observatory.

Three years later, on November 13, 1925, Peltier discovered his first comet, one of 10 comets that bear his name.

DISCOVERING A COMET

The early morning (around 3 am) of May 20, 1990, was partly cloudy but dark as I stepped into my backyard, walked a few steps and pulled the 10 foot square (3 × 3 m) roof off my garden-shed observatory. Although Comet Austin had been visible through binoculars some time earlier in the night, it had set. Besides, I was not planning to look at old comets; I was looking for new ones.

I took the cover off my 16 inch (400 mm) reflector and pointed the telescope toward the east. Then, with eye at the eyepiece and hand on the telescope, I began searching slowly through the eastern sky, moving the telescope one field of view at a time, checking each field for a faint fuzzy object that could be a comet. Finding these faint fuzzies is easy, but most of them turn out to be galaxies, nebulae, or star clusters.

About an hour after I started, the waning crescent Moon rose in the east. Since moonlight floods the sky with enough light to make faint comets disappear, I usually stop hunting when the Moon is up, but on this occasion I did not feel like stopping. I moved the telescope past Alpheratz (Alpha [α] Andromedae), one of the stars in the square of Pegasus, and moved on inside the square.

A few minutes later I stopped moving the telescope and studied the field

COMET LEVY 1990C *moved relative to the stars during this exposure, resulting in trailed star images. The telescope was tracking the nucleus of the comet, which is lost in the larger coma in the photograph.*

of view. There was a fuzzy spot! I knew this part of the sky so well that I suspected this to be the real thing. I contacted the CBAT by electronic mail, with information that included where the object was in the sky and how bright it was.

A day later, when the slowly moving comet had proved me right, the CBAT announced it as Comet Levy (1990c). At the moment of discovery, I had been looking at something that no one had ever seen

before. I felt as though the sky had looked down on me and said, "OK, David, since you've been watching faithfully for so long, here is something special."

Within a week, several observers had reported accurate positions so that the comet's orbit could be determined. The news was good: the comet was expected to get bright enough to be visible to the naked eye. By the end of June, Comet Levy was an easy target through binoculars, and it continued to brighten steadily as the summer progressed. It was at its best in early August. I remember setting up my telescope at a dark roadside location. I was about to turn the scope toward the comet when I looked up and saw the Summer Triangle of Vega, Deneb, and Altair high in the sky. I had first seen those same three stars thirty years earlier, when I was starting as a skywatcher. But on this August night the view was different, for in the middle of that triangle was bright Comet Levy. It was a night I'll never forget.

A LOG BOOK *is an important part of the observer's equipment. Here, David Levy's log book records comet observations and viewing with his friends.*

COMET IKEYA SEKI, *named after two Japanese amateur astronomers, was a spectactular sight in daylight from some parts of the world in 1965. It came so close to the Sun that it was known as a Sungrazing comet.*

associations will supply detailed comparison charts and guidance to the amateur wanting to become serious about variable stars.

RECORDING YOUR OBSERVATIONS

Keeping an observing diary or a log book is perhaps the best way to remember the experiences you have had with your telescope. If you wish to study variables seriously, search for comets, or just take photographs, you will probably find the systematic approach of keeping a diary extremely useful.

OBSERVING A VARIABLE STAR

Observing variable stars is perhaps the easiest really productive work the amateur can undertake, since binoculars are all that are needed. To learn the art, you cannot do better than follow the example of famed variable star observer, Leslie Peltier (see p.72).

For each of the 132,000 variable star observations that Peltier made during his lifetime, he followed a set procedure. After finding the variable with one of his telescopes, he would choose two nearby stars, one a bit brighter than the variable, and another that was fainter. A star chart made especially for the variable would have these "comparison stars" plotted, together with their magnitudes, which might be, say, 8.2 and 8.8. Then by interpolating, he would estimate the brightness of the variable star to be, say, 8.6.

For over 60 years Peltier reported his observations to the American Association of Variable Star Observers (AAVSO). Headquartered in Cambridge, Massachusetts, the AAVSO has followed the behavior of more than 1,000 variable stars since 1911, using reports from amateurs from all over the world. Professional astronomers recognize the AAVSO as a reliable and valuable resource.

Associated programs are run by the British Astronomical Association (BAA) and the Royal Astronomical Society of New Zealand (RASNZ). All three

TEACHING THE TEACHERS

Larry Lebofsky is a planetary scientist who has had a prodigious career. He discovered Larissa, a moon circling Neptune; the existence of water on certain asteroids; and Asteroid 3439 has been named in his honor. But he now tends to use binoculars rather than large telescopes, and spends more time with teachers than with research astronomers.

This new dimension to his career arose from a visit that his wife, Nancy, made to their daughter's second grade classroom. Miranda's teacher had prepared a lesson about the phases of the Moon—first quarter, full, and "gibbons". Lebofsky began wondering why he was studying asteroids when teachers appeared not to know what a gibbous Moon was, so he and Nancy joined forces in building a program that introduces teachers of young children to ways of teaching astronomy. Their Astronomy-Related Teacher Inservice Training (ARTIST) workshops are now held in schools throughout southern Arizona.

OBSERVATORIES *for the amateur come in all shapes and sizes, from the classic dome (below) to the "portable observatory" that is a modified tent (left).*

BACKYARD OBSERVATORIES

Should you become serious about your observing, you may find it useful to build an observatory in your back-yard—somewhere you can keep your telescope set up and ready for use.

A number of companies sell prefabricated structures, ranging from modified tents to simple metal or fiberglass observatory buildings. If you're inclined to be more ambitious, you could even buy a ready-made dome.

You may prefer to make a sliding-roof observatory out of a garden shed, providing you know your way around a hammer and screwdriver. With such a design, when the roof is slid back you will be able to swing your telescope around to see the full sky rather than just a section of it.

If you have a place set up solely for looking at the sky, you will be in the best possible position to begin a serious observing program.

STAR PARTY *One of the benefits of joining an astronomy club is the chance to go along to observing nights. You can then talk to other amateur astronomers about their telescopes and compare the views that they produce.*

JOINING AN ASTRONOMY CLUB

After skywatching on your own for a while, you might find it helpful to join a group of people who are also fascinated by the night sky.

Most major cities have astronomy clubs and their details are given in both *Astronomy* and *Sky & Telescope* magazines, or local astronomy publications. Joining a club almost guarantees that you will have all the help you need to get your telescope working properly. If you are thinking of buying a telescope, you could attend a club viewing night and try the various types that members own. Should you be interested in serious observing, you may find like-minded people who will help you get started with obs-erving variables or patrolling for novae outbursts.

LARGE TELESCOPES

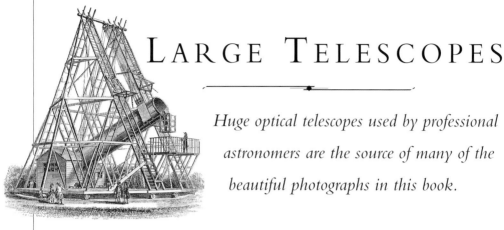

Huge optical telescopes used by professional astronomers are the source of many of the beautiful photographs in this book.

The larger a telescope is, the more light it can gather and the further it can peer into space. George Ellery Hale, born in Chicago in 1868, was well aware of this: "More light!" he would demand. A gifted astronomer, Hale is central to the story of telescope making in the first half of the twentieth century.

At the turn of the century, he persuaded Charles Yerkes, a Chicago businessman, to underwrite a 40 inch (1 m) refracting telescope on the

THE REFLECTING TELESCOPE
(above), built by William Herschel in 1789, had a 48 inch (1.2 m) mirror.

MAUNA KEA *in Hawaii is home to the 400 inch (10 m) Keck telescopes.*

shore of Williams Bay in Wisconsin. One of the world's most beautiful telescopes, the Yerkes telescope is still the largest operating refractor. Hale then arranged funding for the 60 inch (1.5 m) and later the 100 inch (2.5 m) reflector on California's Mount Wilson. But the achievement for which he is most famous is the 200 inch (5 m) reflector on Mount Palomar, California. First used in 1948, it carries his name.

A group of 150 inch (4 m) telescopes in Australia, Chile, and Arizona was completed in the early 1970s, but now a new era of telescope building has begun. The first of two 400 inch (10 m) Keck telescopes has recently been

completed on top of Mauna Kea in Hawaii and a variety of other 300 to 400 inch (8 to 10 m) telescopes are planned. All will employ new mirror construction technologies, simple altazimuth (rather than equatorial) mounts, and extensive computer controls.

OTHER EYES ON THE SKY

Although a small number of amateurs have ventured into the domain of radio astronomy, amateurs are largely restricted by technology to using visible light to scan the universe.

However, since professional astronomy has blossomed beyond this limited view of the universe, it would be

THE ANGLO-AUSTRALIAN TELESCOPE *is one of several 150 inch (4 m) equatorial-mounted reflectors completed in the 1970s.*

RADIO MAPS *are the "pictures" produced by radio telescopes. This is a contour map of the Einstein Ring.*

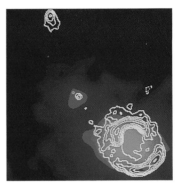

emiss not to acknowledge the growing importance of observing forms of radiation other than light, large amounts of which never penetrate Earth's atmosphere.

RADIO ASTRONOMY

Radio astronomy is the study of celestial bodies by means of the radio waves that they emit and absorb naturally. It has been invaluable in enabling us to discover the nature and shape of the Milky Way and other galaxies. The majority of radio telescopes follow the lead that was set by Grote Reber in the back-yard of his home in 1940, employing large metal parabolic dishes—mirrors which focus the faint radio signals that are received. Often arrays of dishes together perform as one large telescope.

THE INFRARED SKY

Between the visible and radio domains lies the infrared (IR) regime, some of which is accessible from the ground. In the IR, astronomers probe the depth of gas clouds and the center of galaxies using more or less conventional looking telescopes.

ULTRAVIOLET, X-RAYS, AND GAMMA RAYS

Radiation of higher energy than visible light needs to be observed from rockets or satellites. Conventional telescopes are used to detect ultraviolet (UV) and low-energy X-rays, whereas less refined imaging devices are needed to study the exotic and energetic sources of high energy X-rays and gamma rays.

THE VERY LARGE ARRAY (VLA) *in New Mexico uses 80 foot (25 m) diameter dishes to mimic a radio telescope up to 26 miles (42 km) across.*

TELESCOPES IN SPACE

Using an optical telescope on Earth is rather like trying to look at the cosmos from the bottom of a lake. Seeing effects produced by turbulence in the Earth's atmosphere blur the images, and prevent the larger telescopes, in particular, from achieving their resolution goals.

Modern techniques of adaptive optics lessen this impact by measuring and removing the effect, but since the first space flight astronomers have dreamed of putting a large telescope into space above the seeing.

Several small telescopes in space served to whet their appetites, culminating in the launch of the Hubble Space Telescope (HST) into orbit about the Earth in 1990.

After a spectacular repair mission in December 1993, the HST seems set for a productive lifetime exploiting its excellent resolution and working in parallel with larger telescopes on the ground.

THE HST *deployed from the shuttle Discovery in 1990*

CHAPTER FOUR
UNDERSTANDING
the CHANGING SKY

While the sky can seem a bewildering place,
its many marvels and graceful cycles and motions
are well within our grasp.

MAPPING *the* STARS

Although we no longer believe that the Earth is at

the center of the universe, it is a view that can help us understand

why the sky changes from hour to hour and night to night.

BEYOND THE CELESTIAL SPHERE?
This colored woodcut, reminiscent of six-teenth-century engravings, illustrates the old belief that the Earth was surrounded by a vast star-studded dome.

P eople once believed that the stars were attached to a huge sphere that rotated around the Earth once a day. Today we know that the stars and all the other heavenly bodies lie at different distances from us and that they only appear to move around the Earth (from east to west) because the Earth itself is spinning on its own axis (from west to east). Nonetheless, when

describing the positions of stars and the ways in which they appear to rise and set, it is useful to imagine Earth being enveloped by a vast celestial sphere.

EARTH AND THE CELESTIAL SPHERE

To describe the positions of objects in the sky, astronomers developed a grid system of reference points and lines similar to the one used to determine locations on Earth. Here is where the concept of a celestial sphere is of value to us.

Let us imagine that Earth is a balloon that can be inflated until it fills this imaginary celestial sphere. If the most important points and lines on Earth's surface were somehow transferred onto the celestial sphere, the celestial equator would correspond to the Earth's equator, the celestial poles would correspond to Earth's rotational north and south poles, and the lines of longitude and latitude would translate as celestial coordinates called right ascension

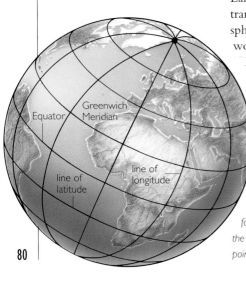

LATITUDE AND LONGITUDE
form a grid on the Earth, defined by the north and south poles, by which any point may be uniquely located.

and declination. Tilted at an angle to the celestial equator would be the ecliptic, the apparent path of the Sun around the celestial sphere.

THE EQUATORIAL COORDINATE SYSTEM

Declination in the sky is measured in the same way that latitude on Earth is measured: in degrees, minutes, and seconds north and south of the equator—in this case, the celestial equator. It increases from 0 degrees on the equator to 90 degrees at the poles. North of the celestial equator declination is listed as positive (+), while south is negative (-).

THE GREENWICH MERIDIAN *is marked by a line at the Old Royal Greenwich Observatory, near London. Here you can stand with one foot in the Eastern and Western Hemispheres.*

THE CELESTIAL SPHERE *and its system of "equatorial" coordinates mimic the surface of the Earth. The important extra feature is the ecliptic—the apparent path of the Sun across the background of the celestial sphere.*

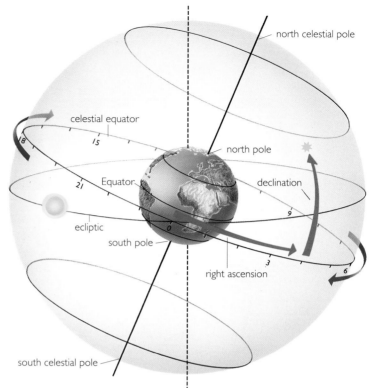

For example, Polaris has a declination of just over +89 degrees; Rigel (in Orion) lies at -8 degrees and 15 minutes.

The celestial equivalent of longitude on Earth is called right ascension (abbreviated to RA), and is measured not in degrees, as longitude is, but in units of time: hours, minutes, and seconds (where 24 hours equals 360 degrees and one hour of RA equals 15 degrees of arc). The right ascension of Polaris is thus referred to as 1 hour, 49 minutes; and that of Rigel 5 hours and 12 minutes. The zero (0) line of RA (the celestial equivalent to Earth's Greenwich meridian of longitude) passes through the point where the Sun crosses the celestial equator at the first moment of spring in the northern hemisphere—the vernal equinox. Hours in right ascension are measured eastward from this point, until we reach 23 hours, 59 minutes. One minute later, of course, and we find ourselves back at 0 hours.

POSITIONS IN THE SKY

The altazimuth coordinate system is especially useful if you are trying to explain to a fellow observer where an object is in the sky at any particular moment. This system is mainly used for navigation, but it also enables skywatchers to specify the position of a celestial body with respect to their horizon and at a particular time using coordinates called altitude and azimuth. These figures differ for each observer, depending on his or her position on Earth and the time that the observation is made.

ALTITUDE (ALSO KNOWN AS ELEVATION) This is the angle above the observer's horizon. The point directly overhead, at 90 degrees, is known as the zenith.

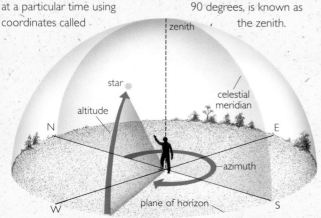

AZIMUTH This is the angle measured clockwise from north along the horizon to the point on the horizon that lies beneath the star. Thus, N = 0 degrees or 360 degrees; E = 90 degrees; S = 180 degrees; and W = 270 degrees.

MERIDIAN This is an imaginary great circle that passes through your zenith from north to south, dividing the sky in two: the eastern and the western halves. It is important to be aware of this line because when an object crosses it, it's as high in the sky as it's going to get.

The Sun crosses the line of the meridian around noon every day. We say that the Sun, or any star, culminates when it crosses the meridian.

A SPINNING EARTH

Every hour of every day, a panorama unfolds above our heads,
and all because the Earth is spinning.

Our planet Earth is a roughly spherical globe which revolves around its own axis, an imaginary line that runs through the center of the Earth from the north pole to the south pole. If it were not for the stars way beyond Earth's atmosphere, it would be difficult to prove that this motion actually takes place, since everything rotates along with us—the land, air, oceans—all of our immediate natural environment.

A SHIFTING HORIZON

As Earth rotates from west to east, our perception is that the stars, the Sun, and all other heavenly objects move around us in the opposite direction—from east to west. The apparent motion of the stars in the sky, however, will depend on where you happen to be on Earth. Once again, the concept of the celestial sphere (introduced on p. 80) can help us explain what stars we can see

and how they appear to move with respect to our particular viewing location. The latitude of the observation point is the key: a skywatcher in Canada sees the stars with much the same perspective as someone in central Europe or in northern Japan; the same is true for people in Buenos Aires, Cape Town, and Adelaide. See illustration below.

At the North Pole, 90° N

For this observer, the celestial north pole, marked by the star Polaris, corresponds with the zenith—the point in the sky directly overhead. From this position on Earth, the celestial equator is parallel with the horizon, and because the stars move along a path parallel to this horizon, only the stars in the northern half of the celestial sphere are ever visible.

At the Equator, 0° From this
latitude, an observer can see all the stars of the sky. Here, the

STAR TRAIL PHOTOGRAPH *showing circumpolar stars arcing around the north celestial pole and never setting*

celestial equator rises up in an imaginary vertical line from the horizon and runs through the zenith. The north and south celestial poles lie exactly on the horizon. The stars in the east rise straight up and sink straight down again below the horizon in the west.

At 40° S Latitude For all
locations lying between these extreme latitudes, there is a part of the sky that always remains invisible—that which surrounds the celestial pole of the opposite hemisphere. On the other hand, the area of the sky close to the visible pole remains in view all the time. Stars here never set, but seem to circle around the celestial pole. These are called circumpolar stars, and the closer you are to the pole, the more circumpolar stars there are.

AT THE NORTH POLE, 90° N

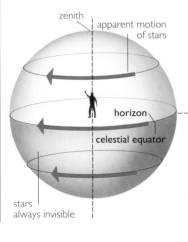

zenith
apparent motion of stars
horizon
celestial equator
stars always invisible

AT THE EQUATOR, 0°

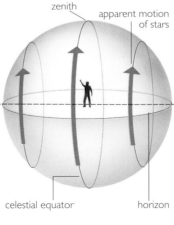

zenith
apparent motion of stars
celestial equator
horizon

AT 40° S LATITUDE

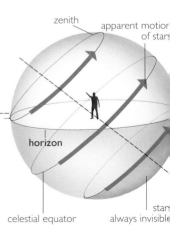

zenith
apparent motion of stars
horizon
celestial equator
stars always invisible

To persons standing alone on a hill
during a clear midnight such as
this, the roll of the world eastward
is almost a palpable movement.

Far from the Madding Crowd,
THOMAS HARDY (1840–1928),
English poet and novelist

TIME *and the* SEASONS

While the tilted Earth spins around its own axis, it is also tracing out a path around the Sun. These two motions, and where we are on Earth, determine what we see in the sky at a given time.

THE SEASONS *personified, in this painting by Walter Crane (1845–1915)*

"**W**hy is it hotter in summer than in winter?" In a recent film called *A Private Universe*, this question was posed to a group of Harvard graduates. Most had no idea of the correct answer—they said that it was because the Earth is closer to the Sun in summer.

SEASONS *on Earth are caused by the tip of the Earth's axis, not by changing distance from the Sun.*

THE REASON FOR THE SEASONS

The Earth orbits the Sun at an average distance of 93 million miles (150 million km). But because Earth's orbit is slightly elliptical the Earth is, just as the Harvard graduates stated, sometimes nearer to the Sun than at other times. But why we experience different seasons has to do with the way the Earth is tilted relative to the plane of its orbit around the

Sun, rather than the distance between them at different times of the year.

From Earth, the Sun is seen projected against the remote background of the celestial sphere. This apparent path of the Sun around the sky in the course of a year is known as the ecliptic. (In the Starfinder Charts in Chapter 5, the ecliptic is marked with a dotted line.) Since most of the planets revolve around the Sun in more or less a flat plane, viewed from Earth the paths of the other planets across the sky tend to stay fairly close to the ecliptic. Knowing this, ancient people attached great import-ance to this band of constella-tions on the ecliptic, known as the zodiac.

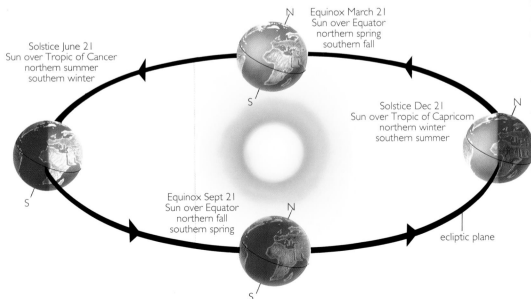

Solstice June 21
Sun over Tropic of Cancer
northern summer
southern winter

Equinox March 21
Sun over Equator
northern spring
southern fall

Solstice Dec 21
Sun over Tropic of Capricorn
northern winter
southern summer

Equinox Sept 21
Sun over Equator
northern fall
southern spring

ecliptic plane

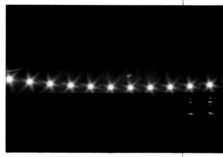

However, the Earth's axis is not perpendicular to the ecliptic—rather, it is tipped at an angle of 23½ degrees. The ecliptic, therefore, is tipped with respect to the celestial equator by the same angle. Because of this, Northern and Southern Hemispheres experience opposite seasons.

In June, the Southern Hemisphere is in winter because it is leaning away from the Sun, while the Northern Hemisphere is experiencing midsummer. The Sun's rays no longer reach the South Pole, and for half the year there the Sun will fail to rise, resulting in continuous night.

Six months later, in December, the Earth will have gone half-way round the Sun. The Northern Hemisphere is now in midwinter, while the Southern Hemisphere is in midsummer. At the South Pole this time, the Sun will remain above the horizon for six months, resulting in a phenomenon referred to as midnight Sun. In March and September, however, both hemispheres have an equal share of day and night.

DAY AND NIGHT

The lengthiness of summer days and the corresponding brevity of winter days increases as you move to higher latitudes. During winter, the Sun is low in the sky, which means that the days are short and the nights long. Also, the Sun's rays must pass through a greater thickness of the atmosphere; some of the heat is absorbed, and the low angle from which the rays come means that they are more spread out.

In summer, on the other hand, the Sun passes high across the sky as it rises and sets each day, yielding longer days than in winter. Also, because the Sun is high during this season, the Sun's rays are less spread out and it heats the Earth more directly.

THE SUN'S MOTION ACROSS THE SKY

Midway between sunrise and sunset the Sun reaches its highest point in the sky around noon. If, over the course of a year, you observed the shadow cast by a stick on a flat piece of ground, you would graphically see how the Sun's height in the sky at noon varies with the passing of the seasons (see illustration below).

During summer, the shadow is shortest at noon on the solstice, the day with the longest daylight hours. Here, the Sun reaches its highest point in the sky for the year. In winter, the Sun traces its lowest path across the sky on the day of the "winter" solstice. It is during this time that the Sun casts the longest shadows and gives us the day with the fewest daylight hours.

PATH OF SUN ACROSS THE SKY *at different times of the year, as viewed by a Northern Hemisphere observer. A Southern skywatcher would look north, instead of south, to see the same effects.*

Summer Solstice

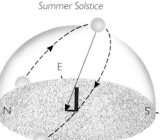

Equinox

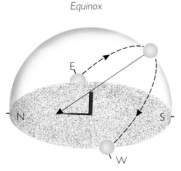

Winter Solstice

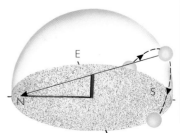

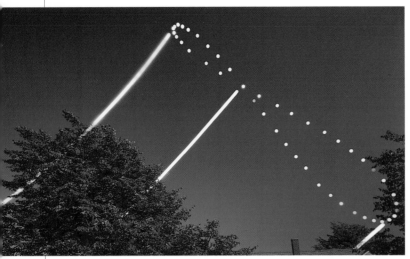

AN ANALEMMA *is the unusual "figure 8" symbol found in various forms on accurate sun dials. This photo illustrates its origin by recording the position of the Sun in the eastern US sky at 8.30 am every 10 days. The Sun's different positions are caused by the tilt of the ecliptic relative to the celestial equator, plus small changes in the Sun's daily speed across the sky.*

At noon on the first days of spring and fall, the length of the shadow cast by the stick is midway between its summer minimum and winter maximum. These days are the vernal and autumnal equinoxes (around March 21 and September 21 for Northern Hemisphere spring and fall, respectively), where day and night are of equal duration.

SOLAR AND SIDEREAL TIME

The Sun's daily motion determines the most basic time cycle—day and night. Our day of 24 hours is how long it takes the Earth to complete one rotation around its axis relative to the Sun. Our clocks indicate time on this basis, and we refer to this as "mean solar time".

We might expect the celestial sphere to rotate around us in the same period, but it does not. Instead, throughout the year as the Earth moves along its path around the Sun, the stars appear to rise about 4 minutes earlier each night—in other words, an observer will see a particular star at the same position 4 minutes earlier on successive nights. Therefore, relative to the stars, the celestial sphere rotates once around the Earth in 23 hours and 56 minutes. Because (for our purposes) the stars are essentially fixed in space, this is the Earth's true period of rotation and is the basis of what we call "sidereal time", the time astronomers refer to when determining an object's position in the sky.

TIME ZONES

All the times given in this book are in mean solar time. We actually use a system of standard times which divides the 360 degrees of longitude around the Earth into 24 time zones, each 15 degrees wide. Because the difference between solar and standard times is usually small, where we are in longitude makes little difference to what we see in the sky, but where we are in latitude does (see p. 105).

When we say, for instance, Orion is on the meridian at 10 pm on Jan 10, this is true for anywhere in the world. (In the case of Orion this is *literally* true, since it spans the celestial equator and can be seen, at

ORION, *as seen looking south in winter from Edinburgh, Scotland. These two views are at the same time of night but two weeks apart, illustrating the slow motion of the sky during the year.*

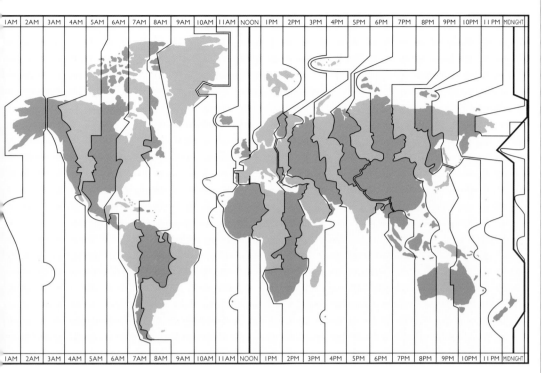

least in part, from anywhere on Earth.) The 10 pm refers to local time.

But what happens when someone in San Francisco wants to compare notes about an eclipsing variable star in Orion with someone in New York City? Orion will indeed be on the meridian at 10 pm local time on January 10 for both of them, but the San Francisco skywatcher will see this 3 hours later in fact! If

they both want to see the star at the same moment as it eclipses, they will need to allow for the time difference in making the observation—say, 10 pm in New York City and 7 pm in San Francisco.

An easier way is to adopt a time system that everyone can use: Universal Time (UT) is essentially Greenwich Mean Time (GMT), which is local time at Greenwich. So now, our two observers can both agree, without confusion, that the eclipse begins at 1 hour UT on Jan 10 (the next day!) because that's the time people in Greenwich will see it.

This simple picture, however, is complicated slightly by "daylight saving" or "summer" time (DST), when we move our local time ahead by 1 hour in order to keep the Sun from rising overly early in summer. Benjamin Franklin proposed this feature in 1784 to save on the cost of lighting.

PRECESSION

Does the Earth's axis always point in the same direction in space? Actually, it does not. Because of tidal effects of the Sun and Moon, the Earth "wobbles" like a spinning top, causing the direction of the vernal equinox to shift in the sky. This wobble is called precession. The sky's entire coordinate system shifts along with that wobble over a leisurely 26,000 years. Therefore, whenever we say that an object has a particular right ascension and declination, we must specify

a precise moment in time, like the year 2000, to define our celestial coordinates. That moment is called the epoch. The vernal equinox is now in Pisces, but over the full wobble it will move through all the signs in the zodiac. The celestial poles move too, so Polaris, our northern pole star, will be of as little use to us a thousand years from now, as it was to the ancient Egyptians.

Sun

celestial pole in AD 14,000

Polaris

equatorial bulge

N

S

TIME ZONES *(above) are used to keep countries, states, or provinces on a common time. As a result, some locations differ by many minutes between true local time and standard time.*

NAMING *the* STARS

Over thousands of years, people of countless cultures have given the stars names and projected their dreams onto them.

Coming mainly from the Arabs and Greeks, the names we use for the stars above us are strange and entertaining. One favorite is Zuben El Genubi, the ancient Arabic name given to the brightest star in Libra. Both the Greeks and the Arabs saw this star as representing nearby Scorpius's southern claw, which is what this splendid name means. The star marking the old northern claw is called Zuben Eschamali.

Betelgeuse is another strange name. Some older texts say, incorrectly, that this should be pronounced "BEETLE-juice", perhaps leading some to wonder why such a majestic star should be diminished to liquid refreshment for bugs. "BET-el-jooze" is the correct pronounciation, which means "house of the twins" in

PTOLEMY *(left) observing the Moon with Urania, who was the Greeks' patron of astronomy*

BAYER'S URANOMETRIA *(1603) was the first serious star atlas, with one map devoted to each of the 48 traditional constellations. This illustration shows Centaurus, the Centaur, Lupus, the Wolf, and Crux, the Southern Cross.*

Arabic, because the Arabs regarded the star as part of the neighboring constellation of

JOHN FLAMSTEED *was the first Astronomer Royal of England and the first director of the Royal Greenwich Observatory. His great star catalogue,* Historia Coelestis Britannica *(1725) was published after his death.*

Gemini. It now marks Orion's shoulder.

BAYER LETTERS AND FLAMSTEED NUMBERS

At the start of the seventeenth century, the German astronomer Johann Bayer set down the positions and magnitudes of all the known stars in an organized way. His star atlas *Uranometria*, published in 1603, labelled many of the stars visible to the naked eye using a system based on the Greek alphabet, whereby the brightest star was generally called alpha (α), the next brightest beta (β), and so on. Betelgeuse is therefore also called Alpha (α) Orionis, Orionis being

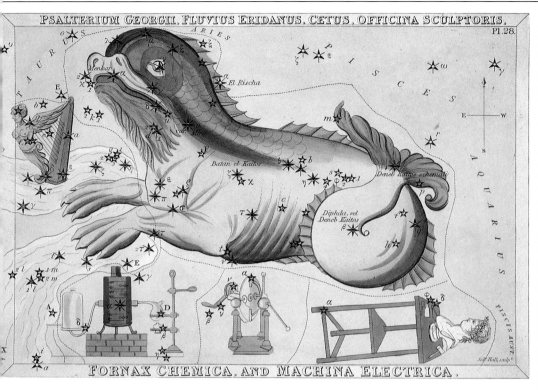

FORNAX CHEMICA, AND MACHINA ELECTRICA.

the genitive form of the constellation name.

Early in the eighteenth century, the Reverend John Flamsteed, appointed by King Charles II of England as the first Astronomer Royal, extended the list of "named" stars by adding Arabic numbers and covering many fainter stars. In his system, the stars are numbered west to east across a constellation, so Betelgeuse is also 58 Orionis.

The fainter a star is, the less interesting is its name. For example, near Zuben El Genubi in Libra is Flamsteed's 5 Librae, best seen through binoculars, and the Smithsonian Astrophysical Observatory's SAO 158846, a magnitude 9.2 star too faint to be seen without a telescope. The SAO Catalogue, which includes 258,997 stars, is another listing often encountered by amateur astronomers. Another bigger and more recent list is the Hubble Space Telescope (HST) Guide Star Catalogue which includes some 19 million stars! There are also many catalogues of star clusters, nebulae, and

URANIA'S MIRROR *(1825) was a set of cards depicting the constellations. Cetus, the Sea Monster, is seen here with some surrounding constellations, including the obsolete Psalterium Georgii and Machina Electrica.*

galaxies. The Messier Catalogue, however, is the most famous, followed by the NGC and IC catalogues (see below).

The stars I know and recognize and even call by name. They are my names, of course. I don't know what others call the stars.

Old woman quoted by Robert Coles in *The Old Ones of New Mexico*

THE MESSIER CATALOGUE

As astronomers grew increasingly familiar with using telescopes and searching the sky, they found more and fainter objects.

From about 1759, French astronomer Charles Messier began listing all the fuzzy objects he encountered in his search for new comets. He listed 103 of these objects in a catalogue, published in 1781, which was to be revised and added to over several years. In its current form the Messier

inventory includes 110 objects—mainly clusters, nebulae, and distant galaxies—spread over most of the sky.

There are also a number of other lists of deep-sky objects, the best known being the New General Catalogue (NGC) of 1888. This, along with its two index catalogues (IC), lists over 13,000 objects. Thus the famous Great Nebula in Orion is known as both M42 and NGC 1976.

89

STAR BRIGHTNESS *and* COLOR

Like lights on a street, stars are fainter the farther away they are, but some stars are supergiants, intrinsically more luminous than mere dwarfs like the Sun and its near neighbors.

The first thing you notice on looking skyward on a clear night is a hodgepodge of stars of varying brightness. How do we describe the brightness of stars and planets?

APPARENT MAGNITUDE

Our system of "magnitudes" dates back to the second century BC when Greek astronomer Hipparchus divided the stars into six brightness groups, the brightest stars being 1st magnitude and the faintest 6th magnitude.

By 1856, Norman Podgson of Radcliffe Observatory had quantified this relationship, making a 1st magnitude star 100 times brighter than the faintest star visible without a telescope. A 2nd magnitude star was then 2.5 times fainter than a 1st magnitude star and a

APPARENT BRIGHTNESS *of stars, like the street lights (below) depends, in part, on their distance. For example, the middle light meter shown in the illustration (right) is twice as far from the light as the front meter and therefore measures only ¼ the brightness. The meter furthest from the light source records only ⅑ of its brightness.*

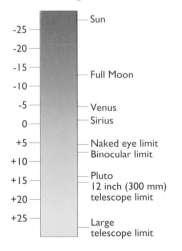

3rd magnitude star, in turn, 2.5 times fainter again.

Podgson defined Polaris, the North Star, as being 2nd magnitude. Most of the stars in the Big Dipper are also about 2nd magnitude. Vega, the brightest star in the Summer Triangle, is 0 magnitude. The brightest star in the sky, Sirius, has a minus magnitude of –1.4. The Moon and Sun are even brighter, of course, but are still accommodated on the scale at –12.6 (full Moon) and –26.8, respectively.

APPARENT MAGNITUDE SCALE
ranges over a factor of around 100 billion in brightness to encompass the brightest stars and the faintest galaxies seen in large telescopes.

Magnitude	
-25	Sun
-20	
-15	
-10	Full Moon
-5	Venus
0	Sirius
+5	Naked eye limit
	Binocular limit
+10	
+15	Pluto
	12 inch (300 mm) telescope limit
+20	
+25	
	Large telescope limit

ABSOLUTE MAGNITUDE

The stars differ in magnitude for two reasons: some are closer to us than others, and some really are brighter than others. In order to describe a star's intrinsic brightness, astronomers have defined absolute magnitude as the apparent magnitude a star would have if it were 10 parsecs or 33 light years from us (see p. 28).

ILLUSION AND REALITY

Comparing apparent and absolute magnitudes is a little like looking at light fixtures on the street. Although all the street lights are the same strength, the ones closer to you appear brighter than the ones down the road. However, the lights in the

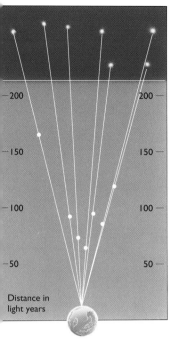

Distance in
light years

ILLUSION VERSUS REALITY *The stars of Ursa Major form the familiar Big Dipper (or Plough) purely by chance since they actually lie at different distances from the Earth.*

STAR COLORS

As a careful look skyward will show, the stars differ not merely in brightness, but also in color. If you look at Antares in Scorpius, Aldebaran in Taurus, or Betelgeuse in Orion, you will see that all these stars are reddish—a subtle red rather than the red of traffic lights. Vega in Lyra and Rigel in Orion are a delicate blue.

A star's color gives us a clue to its nature. Blue signifies that the star is much hotter than the Sun; red means it is cooler (see p. 30).

Star colors are subtle to the eye. In color photographs, however, the colors appear

more obvious, but they are also influenced by the color sensitivity of the film.

The eye itself has a particular color sensitivity—we see yellow and green clearly, but infrared or ultraviolet (UV) light is invisible.

The type of magnitudes we usually encounter are more properly called visual (V) magnitudes since they reflect the brightness of stars as seen by the human eye. Yet some stars are very dim in visible light but bright in, say, the infrared (which we can't see). Our visual magnitude system provides a poor reflection of the true output of such stars.

STAR COLOR *tells us the temperature of the surface of a star. For many stars—the main sequence stars—increasing temperature is usually paralleled by an increase in size.*

20,000° C Blue-white	10,000° C White	6,000° C Yellow	4,500° C Orange	3,000° C Red

...arking lot at the end of the ...treet appear much brighter, ...ven though they are farther ...way, because they are ...ntrinsically more powerful. ...Most reference in this book to ...a star's brightness concerns its ...brightness in the sky as seen ...from Earth—its apparent ...magnitude. Only when we ...wish to establish how ...luminous a star actually is do ...we use absolute magnitudes.

THE TWENTY NEAREST STARS

Name *denotes star has companion	Constellation	Apparent magnitude	Distance (light years)
Proxima Centauri	Centaurus	+11.1	4.24
Alpha Centauri*	Centaurus	-0.27	4.37
Barnard's Star	Ophiuchus	+9.5	6.0
Wolf 359	Leo	+13.6	7.8
Lalande 21185	Ursa Major	+7.6	8.2
Luyten 726-8*	Cetus	+12.3	8.5
Sirius*	Canis Major	-1.42	8.6
Ross 154	Sagittarius	+10.5	9.6
Ross 248	Andromeda	+12.2	10.3
Epsilon Eridani	Eridanus	+3.7	10.6
Ross 128	Virgo	+11.1	10.8
Luyten 789-6	Aquarius	+12.2	11.1
Groombridge 34*	Andromeda	+8.0	11.2
Epsilon Indi	Indus	+4.7	11.3
61 Cygni*	Cygnus	+5.2	11.3
Sigma 2398*	Draco	+8.8	11.4
Tau Ceti	Cetus	+3.5	11.4
Procyon*	Canis Minor	+0.35	11.4
Lacaille 9352	Piscis Austrinus	+7.3	11.5
G 51-15	Cancer	+14.9	11.8

Source: Pasachoff, *Journey Through the Universe* (Saunders College Publishing, 1992)

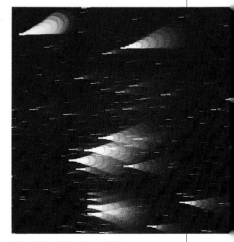

THE COLORS OF STARS *Even photographs do not reveal true star colors because bright stars flood the film with light and appear whitish. Changing camera focus during a star trail exposure overcomes this by spreading out the light to ensure that each star has its color well recorded at some point along its trail.*

THE MOON *and* PLANETS *in* MOTION

Unlike distant stars, the Moon and the planets are not far from us, so we can watch their movements across the sky every night.

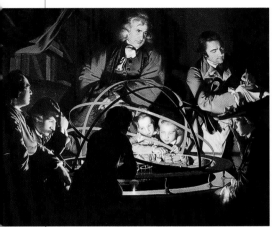

AN ORRERY *was a mechanical device used to illustrate the motion of the planets around the Sun, seen here in a work by British painter Joseph Wright (1734–97). This orrery appears to show the positions of the planets in November 1757.*

The Moon takes a month (a "moonth") to orbit the Earth. As a result of the gravitational pull between the Earth and the Moon, it also takes the same amount of time to rotate on its axis, so it always keeps the same face turned toward us. However, this face is presented about 50 minutes later each night as the Moon moves among the stars, because of its orbit around the Earth.

The Moon also presents a varying amount of its lit face to us during its monthly orbit, because of the changing relationship between it, the Sun, and the Earth. These are known as its phases.

At full phase, it is opposite the Sun from our point of view, so we see the entire side of it that faces us bathed in sunlight. When it is new, the Sun shines on the far side of the Moon, so the sunlit face is hidden from us. At other phases, we see only part of the Moon's sunlit surface. All the time, whether it is lit or not, we are only ever looking at the one face of the Moon.

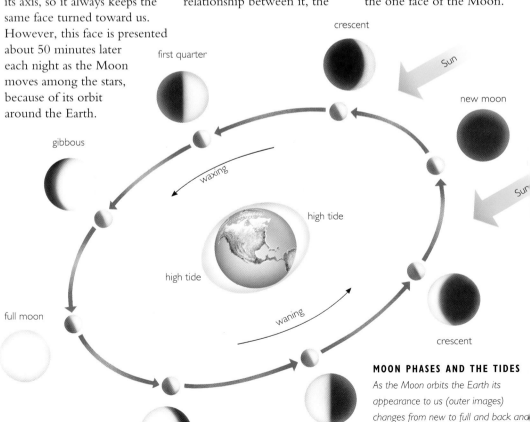

crescent

first quarter

Sun

new moon

gibbous

waxing

high tide

Sun

high tide

full moon

waning

crescent

MOON PHASES AND THE TIDES
As the Moon orbits the Earth its appearance to us (outer images) changes from new to full and back and its gravitational pull causes the tides.

gibbous

last quarter

Here on Earth, the ebb and flow of the tides are a constant reminder of the Moon's gravitational influence (see p. 239).

THE WALTZ OF THE PLANETS

Although they move much more slowly than the Moon, the planets trace elegant paths among the stars that you can easily follow during a season of observing. The farther away a planet is from us, the more slowly it appears to move against the stars at any time.

During the course of some months, you can actually watch a planet being overtaken by the Earth in the same way that you see a car recede as you overtake it on the highway. As you pass the car, it appears to slow down, then move backward, then resume its forward movement after you have gone by.

Moving more slowly around the Sun than we do, Mars, Jupiter, and Saturn all exhibit the same type of motion, moving eastward, appearing to reverse course

as we pass them, and then resuming their eastward trek. This process is called retrograde motion.

THE INNER PLANETS

Mercury and Venus are closer to the Sun than we are, and orbit it more quickly. Their movements appear very different from those of the outer planets.

Venus, for example, will appear close to the Sun early one evening. With each passing week, it will appear to rush away from the Sun, then it will slow down and for a few days it will seem to move very slowly—a time known as greatest elongation. From a maximum of 47 degrees from the Sun, it will then start to move closer again, becoming so close that we cannot see it. Its closest apparent position to the Sun is called conjunction. In fact, at this point it is no closer to the Sun than usual; it is simply almost in line with us and the Sun. After that, it appears in the morning sky again and repeats its performance.

RETROGRADE MOTION *is a characteristic of the outer planets, as the Earth overtakes them in its faster orbit. Here, a retrograde loop of Mars is superimposed on the stars of the constellation Taurus*

Mercury moves in the same way, but it goes through the acts of its drama much faster than Venus, and it never reaches more than 28 degrees from the Sun.

CRESCENT NEPTUNE, *as seen from Voyager 2, illustrates how a planet that is closer to the Sun than the observer can show a crescent phase.*

SOLAR *and* LUNAR ECLIPSES

If there were to be a list of the "top ten"

inspiring sights in the world, a total eclipse

of the Sun would surely be included.

ECLIPSE IN 1688 *is observed by projection through a telescope by Jesuit priests attending the King of Siam (former name of Thailand)*

Everyone should try to see at least one total eclipse of the Sun. Seeing the Moon come exactly between the Sun and the Earth, and slowly consume the Sun, is one of the marvels of being alive. For most of us, such an eclipse will not just happen—we will need to seek it out by traveling to a place where the narrow shadow of the Moon strikes the Earth. As the Earth rotates, the shadow traces out a dark path up to about 200 miles (300 km) wide but thousands of miles long. Within this path, any one location will be dark for a few minutes only.

From the time the Moon takes its first tentative bite out of the Sun, it marches forward in an ominous embrace. The increasing darkness of the sky may not be apparent at first, as if a light cloud has covered the Sun, but when you look at it through your properly filtered telescope (see p. 65), the Sun has become a crescent. By the time four-fifths of the Sun is covered, the darkness is deepening rapidly. The breeze stops, the temperature falls, and even the wildlife quietens down in anticipation of some strange nightfall. When you look off toward the horizon you see a black shadow rushing toward you, and when you look toward the Sun again it has disappeared, leaving the softly glowing jewelled crown —the corona—in its place.

After too short a time, one side of the Sun brightens, a flash of light darts out, and the spectacle has come to an end. The Sun is reappearing.

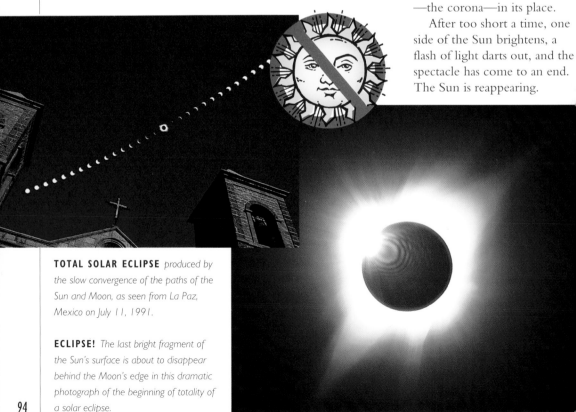

TOTAL SOLAR ECLIPSE *produced by the slow convergence of the paths of the Sun and Moon, as seen from La Paz, Mexico on July 11, 1991.*

ECLIPSE! *The last bright fragment of the Sun's surface is about to disappear behind the Moon's edge in this dramatic photograph of the beginning of totality of a solar eclipse.*

ANNULAR ECLIPSE *photo-graphed just before sunset. The disk of the Moon was clearly too small to cover the Sun on this occasion.*

A RING OF FIRE

If the Moon is near the farthest point of its orbit around the Earth, it may not completely cover the Sun. These are known as annular eclipses, because an annulus, or ring, of brilliant sunlight, looking like a ring of fire, appears around the rim of the Moon, providing it is centered on the Sun's disk.

OBSERVING A SOLAR ECLIPSE

Never look directly at a partially eclipsed Sun, or without a proper solar filter on your telescope (see p. 65). The danger of blinding oneself is high during the partial phase of

an eclipse, because the low level of light allows us to stare at the Sun without squinting. It is safe to look directly at a total eclipse in the few minutes of totality, as none of the Sun's disk is visible—but make sure you turn away the instant the Sun reappears.

A total eclipse of the Sun provides the only occasion on Earth when we can easily see the Sun's hot and tenuous outer atmosphere or corona. Normally the intense glare of the solar surface overwhelms the corona's faint glow. We can also see prominences—the jets of gas that leap from the Sun's surface.

LUNAR ECLIPSES

When the Earth passes between the Moon and the Sun, a lunar eclipse occurs for everyone on the nighttime side of the Earth. The Earth usually casts a coppery, dim light on the Moon, rather than a black shadow, but after major vol-canic eruptions, dust in the Earth's atmosphere darkens the eclipse to a red or brown.

During a total eclipse, the entire Moon is in the shadow of the Earth, often for over an hour. A partial eclipse occurs when only some of the Moon's disk is in shadow.

VENUS ABOUT TO DISAPPEAR! *Occultations occur when the Moon passes in front of a more distant object. Here, Venus is about to be concealed by the crescent Moon, on its journey eastward against the background stars.*

PARTIAL LUNAR ECLIPSE *This image was made by opening the camera shutter for 1/60 th of a second every 20 or 30 minutes during the eclipse.*

SOLAR ECLIPSE

sunlight

Moon
umbra
penumbra
area of total eclipse
area of partial eclipse
Earth

ANNULAR ECLIPSE

sunlight

Moon
umbra
penumbra
area of annular eclipse
Earth

LUNAR ECLIPSE

sunlight

Earth
umbra

Moon
penumbra

ECLIPSES *When the Moon's shadow strikes the Earth's surface, a total (top) or annular (center) eclipse results for observers within a narrow band on Earth. In a lunar eclipse (bottom) the situation is reversed, and the much larger shadow of the Earth easily envelops the Moon.*

"SHOOTING STARS"

What we commonly call "shooting stars"
are not stars at all, but tiny particles sweeping
into the Earth's atmosphere from space.

Meteoroids are tiny particles the size of a grain of sand, usually the residue from comets, that round the Sun in elliptical orbits. If a meteoroid encounters the Earth's upper atmosphere, it vaporizes in an event we call a meteor. If the object is large enough to penetrate the atmosphere and hit the ground, we refer to it as a meteorite.

OBSERVING METEOR SHOWERS

From a dark country sky, one can observe around a dozen meteors per hour any night of the year. But on some nights, depending on where the Earth is in its orbit around the Sun, the Earth comes across streams of them, as it did in spectacular

METEORITE *(above, left) about 10 inches (250 mm) across and weighing 4 lb (8 kg). This was found in Antarctica but probably came from Mars!*

LEONID METEORS *(above) A false color image of meteor trails from the 1966 meteor storm, shown against a background of stars*

fashion during the Leonid meteor storm in 1966, when some observers saw 40 meteors every second.

When we look down a railroad, the tracks appear to converge, and it is the same when we observe a meteor shower. Because of the effects of perspective, the meteors appear to radiate from one point in the sky. Most showers are thus named for the constellation in which this point (the radiant) lies.

METEORITES

Meteorites can be spectacular. On April 26, 1803, a shower of thousands of small stones fell over northwestern France, terrifying people. On November 30, 1954, an Alabama woman was inside her home when a small meteorite shattered her roof, bounced off her radio, and hit her. She was not seriously hurt

MAJOR ANNUAL METEOR SHOWERS			
Shower	Date	Hourly	Parent comet
Quadrantids	January 3	40	
Lyrids	April 22	15	Comet Thatcher
Eta Aquarids	May 5	20	Comet Halley
Delta Aquarids	July 28	20	
Perseids	August 12	50	Comet Swift-Tuttle
Orionids	October 22	25	Comet Halley
Taurids	November 3	15	Comet Encke
Leonids	November 17	15	Comet Temple-Tuttle
Geminids	December 14	50	Asteroid 3200 Phaethon
Ursids	December 23	20	Comet Tuttle

Dates can vary slightly. Hourly rate represents the number of meteors you might see under a dark sky when the radiant is near the zenith. Expect to see perhaps half as many more if the shower is strong.

People all over the midwestern United States witnessed a fireball as bright as a quarter Moon on the evening of October 9, 1992. Lighting up the southwestern sky, it took approximately 15 seconds to cross the sky before disappearing in the northwest. A few minutes later, its fate became apparent when Michelle Knapp was startled by a crash outside her home in Peekskill, New York. She found her car's trunk crushed and a football-sized rock lying nearby. She notified authorities, who suspected it was a meteorite when they found that it was very heavy and was still warm.

TYPES OF METEORITES

While it is difficult for an inexperienced person to tell if a rock is a meteorite, there are a few things that you can look for. If the rock has flat surfaces or sharp, angular edges, it will not be a meteorite, nor will it be a meteorite if it is

METEOR CRATER *in Arizona is ¾ mile (1.2 km) across and 650 feet (200 m) deep and was formed by a meteor crashing into the Earth about 50,000 years ago.*

crystalline. A "fused" outer crust is, however, a clue that the rock has plunged through the atmosphere.

A meteorite rich in iron will be attracted to a magnet. If you think you have found a meteorite containing iron, you can perform a "streak test". Clean the sample of rust, then scratch it across an unglazed ceramic tile. If it leaves a grey streak, it is probably just magnetite. If it leaves virtually no streak at all, it might well be a meteorite.

METEOR *captured on film as it flashed across a trail photograph of the constellation Orion*

Today most meteorites are found in searches by scientists or meteorite dealers. The favorite locations are open, unpopulated tracts of land like Australia's Nullarbor Plain. Perhaps most productive of all in recent times has been the ice sheets of Antarctica wherein a record of thousands of years of meteorite falls can be found, locked in pristine condition.

OTHER LIGHTS *in the* SKY

Aurorae, haloes, and rainbows—some of the sky's most wondrous sights—are not far off in space but in the atmosphere close above us.

When charged particles from active areas on the Sun interact with the Earth's atmosphere, a skywatcher may be treated to an aurora. These displays are most commonly visible around the Earth's magnetic poles—in the north as aurora borealis and in the south as aurora australis.

These lights take a variety of forms, the most basic being a greenish glow near the horizon. If the glow becomes stronger, it might form an arc of light, and if rays start to quiver and dance above the arc, the display can be resplendent. As the rays strengthen, they in turn can become shimmering curtains of light. Sometimes, a strong all-night display is capped with a corona of rays directly overhead.

RAINBOWS

Rainbows are the best known and perhaps the most romantic of the lights in the sky. They occur when sunlight is refracted through droplets of water. Each drop acts as a prism,

AN AURORAL CURTAIN *seen above the spruce forests of Alaska. The group of stars just above the trees at the left is the constellation of Delphinus.*

RAINBOW *arcing across farmland near Bordeaux, in France*

with different colors resulting from the light being bent by slightly different amounts.

A rainbow is actually not a bow but forms a complete circle, centered about the point opposite the Sun. Thus, a rainbow will never appear when the Sun is near the zenith.

HALOES

When high cirrus clouds move in, the Sun or the Moon may appear to develop a halo. These haloes arise when sunlight or moonlight refract through lens-like ice crystals in the clouds. Sometimes double haloes develop and haloes form off other haloes.

If the cirrus thickens, patches of light called sundogs, or parhelia, sometimes appear,

nd because the Sun will have
een dimmed by the cloud, it
nd the sundog can both seem
qually bright.

ARTIFICIAL SATELLITES
f an observer from the
ineteenth century were to
eappear tonight, he or she
vould be amazed by the
undreds of moving lights in
he sky. These artificial moons,
aunched into orbit about the
Earth mainly by the United
States and the former Soviet
Union, are best seen in the
our after dusk or before
lawn, when sunlight reflects
ff them but the sky is still
elatively dark.

However, do not expect
he satellite that bounces a TV
ignal to your antenna to be
pparent as a moving point of
ight. Such a satellite will be in
geosynchronous orbit some
2,300 miles (35,700 km)
bove the Earth, meaning that
ts orbital speed matches the
Earth's rotation and it there–

ROCKET TRAIL *from a
Minuteman 3 test has been dispersed
ere by high altitude winds. Ice crystals
roduce the rainbow-like colors.*

A HALO *with a radius of 22 degrees is
the brightest and most common halo
seen. Haloes are caused by hexagonal
ice crystals in high clouds. The reddish
inner rim is a common feature.*

SUN DOGS *(right), or parhelia, are
closely related to 22 degree haloes, often
appearing just outside them.*

fore always remains above the
same place. It will also be well
below naked eye visibility.

**UNIDENTIFIED
FLYING OBJECTS**
Anything that appears to be
flying that cannot be identified
is known as an unidentified
flying object (UFO).

Although some experienced
skywatchers have occasionally
encountered such sights,
most of the widely publicized
UFO reports come from
inexperienced observers.
Venus in the evening sky, or a
bright fireball, are commonly
mistaken for UFOs. For most
of us, the more time we spend
observing the sky and
becoming familiar with its
many features, the less likely
we are to spot a UFO.

CHAPTER FIVE
A GUIDE *to the* SKY

As the earth spins, it reveals an ever-changing view of the sky. These Starfinder charts will guide you through the constellations as they appear and disappear during the night and throughout the seasons.

FINDING YOUR WAY AROUND

Finding constellations can be quite a challenge for the beginner. The trick is to start by identifying some of the brighter stars, and then to "star-hop".

ASTRONOMICAL KNOWLEDGE *is used by navigators in this detail from a work by Cornelis de Baeilleur (1607–71).*

Finding your way around the sky can seem daunting, but it is really no harder than reading a road map, and a good deal more relaxing. When you drive along a highway, you have only a few minutes to match the road signs to the map before you risk missing an important turn. By comparison, the stars do not race by; instead they appear to move gracefully across the sky. And tomorrow night they will appear much the same.

But how do you find your way to a particular place in a sky that is so vast and so full of stars? By using a star at a time as a guide, you can "star-hop" your way around.

FINDING NORTH

For Northern Hemisphere dwellers, the sky provides a bright star close to the north celestial pole which is a good place to start. It also provides a convenient way to find it. Just find the Big Dipper (in Ursa Major), mentally draw a line joining the two stars at the end of the bowl, extend it five times, and you're at Polaris.

This is easy if the Big Dipper is in the sky, which it is from mid-northern latitudes every evening of the year. However, in fall and winter you may not be able to see it unless you have a clear northern horizon. From the southern United States, the Dipper lies below the horizon on winter evenings.

If the Dipper is low, then the W shape of Cassiopeia, on the other side of the pole, will be high. It does not point the way as clearly as the Dipper, but it will give you an idea of the direction of the Pole Star.

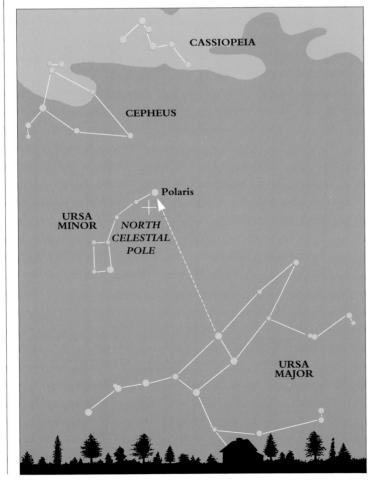

FINDING NORTH *is easy for Northern Hemisphere skywatchers because the bright star Polaris lies very close to the north celestial pole.*

(Labels on diagram: CASSIOPEIA, CEPHEUS, Polaris, URSA MINOR, NORTH CELESTIAL POLE, URSA MAJOR)

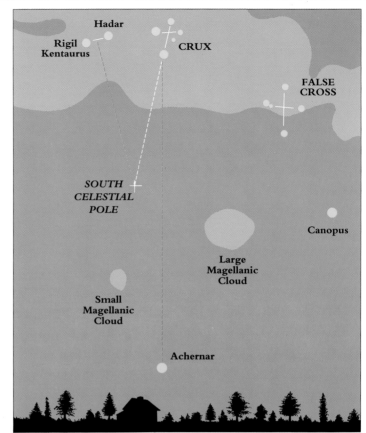

FINDING SOUTH *is made harder by the lack of a bright star near the pole, but other bright stars provide clues to the pole's position.*

FINDING SOUTH

For Southern Hemisphere skywatchers, finding south is not quite so easy. But although the south celestial pole is not marked by a bright star, there are several well-known ways to find the pole. The easiest is to simply extend the long arm of Crux, the Southern Cross, four and a half times to reach close to the pole. The pole is very nearly marked by a star, Sigma (σ) Octantis, but this is too faint really to be useful.

OTHER SIMPLE JOURNEYS

The Big Dipper can lead you to other stars and constellations in the sky. Joining the three stars in the handle forms a curved line or arc. If you extend the line away from the Dipper you can "arc to Arcturus", the brightest star in Boötes, the Herdsman, and then "speed to Spica", in the same direction—the bright star in Virgo. Judging these distances in the sky can be tricky to begin with, but it just takes a little practice.

You can devise your own star-hopping journeys. Start at a bright star and move around the sky, one star at a time, until you reach your destination. Constellations containing a particularly bright star are marked with a special symbol in the constellation charts.

BRIGHT CONSTELLATIONS

In finding your way around the sky, it rapidly becomes apparent that a few constellations are bright and easy to find. In the constellation charts that follow they are rated 1 on the visibility scale and serve as handy jumping-off points for your star-hopping journeys across the sky.

For example, Orion is prominent in the sky from most locations in the first few months of the year. More or less opposite Orion in the sky is Scorpius, which is prominent in the northern summer (southern winter) sky. Other well-known landmarks include the Big Dipper in Ursa Major, the W shape of Cassiopeia, Leo with its sickle, the Great Square of Pegasus, and Crux, the Southern Cross.

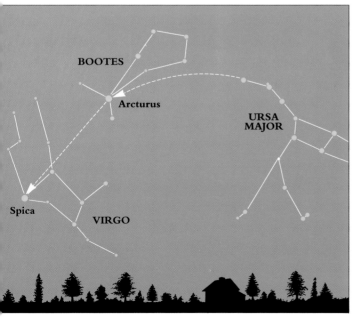

ARC TO ARCTURUS *and speed on to Spica is an expression well known to northern skywatchers learning their way between bright stars in the sky.*

THE STARFINDER CHARTS

The Starfinder charts are your map to the night sky. Begin with the sky charts, and then zoom in on your destination with a constellation chart.

The charts can be used anywhere in the world, but remember that not all constellations will be visible from any given place; for example, constellations in the south polar region of the sky will always remain below the horizon for mid-northern latitude observers, and vice versa (see pp. 80–1).

On the charts, solid lines link some of the major stars of individual constellations. Today, 88 constellations are recognized, with boundaries defined by a commission of the International Astronomical Union in 1930. The lines connecting groups of stars form the configurations that are the basis of the stories associated with the constellations and the naming of them.

THE SKY CHARTS

The bimonthly sky charts will help you find your way amongst the stars. By choosing a chart appropriate to the time and date of observation, you should be able to match a chart with your view of the sky.

There are twelve sky charts in all—six for Northern Hemisphere skywatchers and six for those in the Southern Hemisphere. Each is in two parts on facing pages—the left-hand chart looking north, the right-hand chart looking south—with considerable overlap in the middle. Each is designed to be held with the writing upright.

The size of the dots representing the stars indicate their relative brightness—the larger the dot, the brighter the star. A key to the symbols used appears with each chart. Most of the stars plotted are of magnitude 4.5 or brighter—bright enough to be easily seen with the naked eye in a rural setting.

CHOOSING A SKY CHART

The constellations visible from any given part of the world appear to rotate one-sixth of the way around the sky every two months, so at the same time of night, every two months, we see an additional 60 degrees of sky in the east and lose sight of 60 degrees in the west (except at the poles).

Each chart is valid for specific times on specific days. For instance, Sky Chart 1 is valid for 12 am, Jan 1; 11 pm, Jan 15; 10 pm, Feb 1; and so on, as indicated alongside the chart. Sky Chart 2 is valid four hours later on any of these dates, with the arrows showing the direction in which the sky rotates.

Each sky chart is also valid for other months, but at different times of the night (or day!). Choose a chart which is the closest match to the date and time you want, as indicated in the Sky Chart Tables above. On these tables,

FIND YOUR LATITUDE

SKY CHART TABLE: NORTHERN HEMISPHERE

Local Time	6pm	7pm	8pm	9pm	10pm	11pm	12am	1am	2am	3am	4am	5am	6am
DST	7pm	8pm	9pm	10pm	11pm	12am	1am	2am	3am	4am	5am	6am	7am
January 1			Chart 6				Chart 1				Chart 2		
January 15		Chart 6				Chart 1			Chart 2				
February 1	Chart 6				Chart 1			Chart 2					Chart 3
February 15				Chart 1				Chart 2				Chart 3	
March 1			Chart 1				Chart 2				Chart 3		
March 15		Chart 1				Chart 2			Chart 3				
April 1	Chart 1				Chart 2			Chart 3					Chart 4
April 15				Chart 2				Chart 3				Chart 4	
May 1			Chart 2				Chart 3				Chart 4		
May 15		Chart 2				Chart 3			Chart 4				
June 1	Chart 2				Chart 3			Chart 4					Chart 5
June 15				Chart 3				Chart 4				Chart 5	
July 1			Chart 3				Chart 4				Chart 5		
July 15		Chart 3				Chart 4			Chart 5				
August 1	Chart 3				Chart 4			Chart 5					
August 15				Chart 4				Chart 5				Chart 6	
September 1			Chart 4				Chart 5				Chart 6		
September 15		Chart 4				Chart 5			Chart 6				
October 1	Chart 4				Chart 5			Chart 6					
October 15				Chart 5				Chart 6				Chart 1	
November 1			Chart 5				Chart 6				Chart 1		
November 15		Chart 5				Chart 6			Chart 1				
December 1	Chart 5				Chart 6			Chart 1					
December 15				Chart 6				Chart 1				Chart 2	

SKY CHART TABLE: SOUTHERN HEMISPHERE

Local Time	6pm	7pm	8pm	9pm	10pm	11pm	12am	1am	2am	3am	4am	5am	6am
DST	7pm	8pm	9pm	10pm	11pm	12am	1am	2am	3am	4am	5am	6am	7am
January 1			Chart 12				Chart 7				Chart 8		
January 15		Chart 12				Chart 7			Chart 8				
February 1	Chart 12				Chart 7			Chart 8					Chart 9
February 15				Chart 7				Chart 8				Chart 9	
March 1			Chart 7				Chart 8				Chart 9		
March 15		Chart 7				Chart 8			Chart 9				
April 1	Chart 7				Chart 8			Chart 9					Chart 10
April 15				Chart 8				Chart 9				Chart 10	
May 1			Chart 8				Chart 9				Chart 10		
May 15		Chart 8				Chart 9			Chart 10				
June 1	Chart 8				Chart 9			Chart 10					Chart 11
June 15				Chart 9				Chart 10				Chart 11	
July 1			Chart 9				Chart 10				Chart 11		
July 15		Chart 9				Chart 10			Chart 11				
August 1	Chart 9				Chart 10			Chart 11					
August 15				Chart 10				Chart 11				Chart 12	
September 1			Chart 10				Chart 11				Chart 12		
September 15		Chart 10				Chart 11			Chart 12				
October 1	Char 10				Char 11			Char 12					
October 15				Chart 11				Chart 12				Chart 7	
November 1			Chart 11				Chart 12				Chart 7		
November 15		Chart 11				Chart 12			Chart 7				
December 1	Chart 11				Chart 12			Chart 7					
December 15				Chart 12				Chart 7				Chart 8	

sunset and sunrise are indicated by the edge of the lighter zones, but the sky is not really dark within about an hour and a half of these times. Also watch out for the difference between standard time and daylight saving (summer) time—DST in the charts and tables.

Across the bottom of each sky chart is a series of curved lines representing the horizon for observers at different latitudes. (The world map opposite will help you to determine your latitude.) Any stars lying beneath the horizon line applicable to your location will not be visible. In the upper-middle portion of each map, the zenith (the point directly overhead) for observers at each latitude is marked with a "plus" sign.

The ecliptic (the plane of the Solar System) is marked on each chart by a dotted line. The Moon and planets, if they are above the horizon, will be close to this line. A bright 'star' near the ecliptic, but not on the charts, will almost certainly be a planet. The wavy, pale-colored area that appears on the charts represents the band of the Milky Way.

25 BRIGHTEST STARS ☆

Common name	Constellation name	App. mag.
Sirius (d)	α Canis Majoris	-1.46
Canopus	α Carinae	-0.72
Alpha Centauri (d)	α Centauri	-0.01
Arcturus	α Bootis	-0.04
Vega	α Lyrae	0.03
Capella	α Aurigae	0.08
Rigel	β Orionis	0.12
Procyon	α Canis Minoris	0.8
Achernar	α Eridani	0.46
Hadar (v)	β Centauri	0.66
Betelgeuse (v)	α Orionis	0.70
Altair	α Aquilae	0.77
Aldebaran	α Tauri	0.85
Acrux	α Crucis	0.87
Antares (v)	α Scorpii	0.92
Spica (v)	α Virginis	1.00
Pollux	β Geminorum	1.14
Fomalhaut	α Piscis Austrini	1.16
Deneb	α Cygni	1.25
Beta Crucis (v)	β Crucis	1.28
Regulus	α Leonis	1.35
Adhara	ε Canis Majoris	1.50
Castor (d)	α Geminorum	1.59
Shaula (v)	λ Scorpi	1.62
Bellatrix	χ Orionis	1.64

(d) = double star (v) = variable star

LONGITUDE OF YOUR OBSERVING SITE

Contrary to what you might think, this is not too important. Observers in Tucson and Tel Aviv (despite being in two distinctly separate longitude zones) will see much the same sky at the same *local* time— say, 9 pm at each location— because they share the same latitude zone. Small differences do arise because civil time is based on standard time zones rather than actual local time, but this effect is minor.

STEP-BY-STEP GUIDE TO USING THE SKY CHARTS

1 Before turning to the sky charts themselves, determine the hemisphere and latitude of your location by referring to the world map on p. 104.
2 Use a Sky Chart Table on p. 105 to work out which chart is applicable for your date and time of viewing, then turn to the sky chart.
3 Decide whether you will be facing north or south, and refer to the relevant half of the chart.
4 Look at the horizon lines on the chart and determine which

one is applicable to your latitude; also find the corresponding zenith point (the point directly overhead).
5 Choose two or three of the brightest stars featured (the biggest dots on the chart) and try to find them in the sky, noting where they are relative to the horizon and zenith.
6 Once you've identified one of these stars, try to trace out the constellation of which it is a part, then trace out the patterns of constellations nearby.

Example If you live in San Francisco and you plan to go skywatching at 10 pm (11 pm daylight saving time) on April 1, here's what you should do.
1 Turn to the map on p. 104 to find the latitude of San Francisco, which is approximately 40° N (actually 37°48'). The horizon line and zenith relevant to you on all the Northern Hemisphere sky charts will therefore be those marked 40°N.
2 According to the Northern Hemisphere Sky Chart Table on p. 105, you should turn to Sky Chart 2 on pp. 110–1.
3 On the south-facing half of Sky Chart 2 you will see that the prominent constellation of Orion can be found low in the western sky with the brightest star Sirius near the south-western horizon.
4 If you want to find out more about Orion, turn to its constellation chart on p. 194. Together, the sky charts and constellation charts will guide you through the night skies.

THE CONSTELLATION CHARTS

Once you're ready to zero in on a particular constellation, turn to the constellation charts. These are presented in alphabetical order, and with north at the top and east to the *left*— different from maps of the Earth, but necessary to match our view of the sky.

Within the constellation, stars down to magnitude 6.5 are shown; those which fall *outside* the constellation are star down to magnitude 5.5. All the stars marked are therefore visible to the naked eye under dark skies, but in towns or cities, binoculars may be needed to pick out the fainter ones. (As a general rule-of-thumb, the appropriate magnitude limits are: cities—2 or 3, suburbs—4, distant suburbs—4.5 or 5, rural—5 to 6.5.). In addition to the naked-eye stars, the charts mark the locations of other important objects such as star clusters, nebulae, and galaxies; most of these require optical aids to be seen. Any deep-sky objects of particular interest are included, down to around magnitude 11.

A key to the symbols used on the charts appears below. Note also that many stars are named with letters of the Greek alphabet. These are shown clearly on the charts. Each chart is accompanied by descriptions of the main objects of interest to the amateur observer, with an indication of the type of instrument needed to see them.

KEY TO SYMBOLS

Magnitudes   ● -1   ● 0   ● 1   ● 2   ● 3   • 4   · 5   · 6 and under

Double stars ●-●   Variable stars ◌ ○   Open clusters ◌ ◌

Globular clusters ⊕ ⊕   Planetary nebulae ◇ ◇

Diffuse nebulae ▱ □ ▫   Galaxies ◯ ◯   Quasar ◌

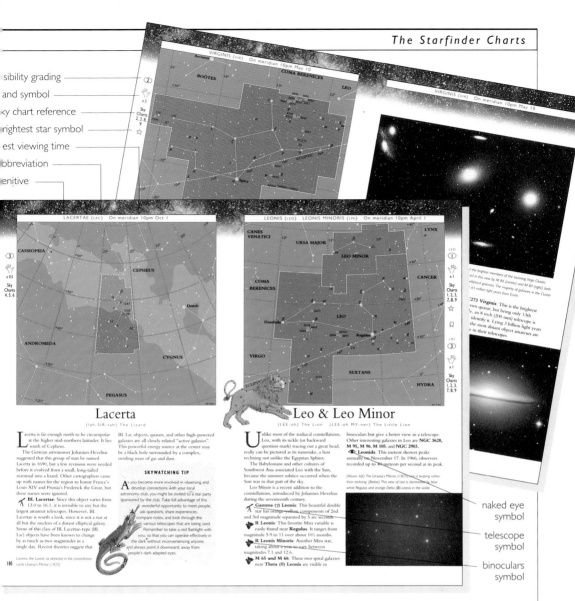

sibility grading

and symbol

sky chart reference

brightest star symbol

best viewing time

abbreviation

genitive

VIRGINIS (VIR) On meridian 10pm May 10

VIRGINIS (VIR) On meridian 10pm May 10

LACERTAE (LAC) On meridian 10pm Oct 1

LEONIS (LEO) LEONIS MINORIS (LMI) On meridian 10pm April 1

Lacerta
(lah-SIR-tah) The Lizard

Lacerta is far enough north to be circumpolar at the higher mid-northern latitudes. It lies south of Cepheus.

The German astronomer Johannes Hevelius suggested that this group of stars be named Lacerta in 1690, but a few revisions were needed before it evolved from a small, long-tailed mammal into a lizard. Other cartographers came up with names for the region to honor France's Louis XIV and Prussia's Frederick the Great, but these names were ignored.

BL Lacertae: Since this object varies from 13.0 to 16.1, it is invisible to any but the largest amateur telescopes. However, BL Lacertae is worth a look, since it is not a star at all but the nucleus of a distant elliptical galaxy. Some of this class of BL (BL Lac) objects have been known to change by as much as two magnitudes in a single day. Recent theories suggest that

BL Lac objects, quasars, and other high-powered galaxies are all closely related "active galaxies". This powerful energy source at the center may be a black hole surrounded by a complex, swirling mass of gas and dust.

SKYWATCHING TIP

As you become more involved in observing and develop connections with your local astronomy club, you might be invited to a star party sponsored by the club. Take full advantage of this wonderful opportunity to meet people, ask questions, share experiences, compare notes, and look through the various telescopes that are being used. Remember to take a red flashlight with you, so that you can operate effectively in the dark without inconveniencing anyone, and always point it downward, away from people's dark adapted eyes.

Lacerta, the Lizard, as depicted in the constellation cards Urania's Mirror (c.1825)

Leo & Leo Minor
(LEE-oh) The Lion (LEE-oh MY-ner) The Little Lion

Unlike most of the zodiacal constellations, Leo, with its sickle (or backward question mark) tracing out a great head, really can be pictured as its namesake, a lion reclining not unlike the Egyptian Sphinx.

The Babylonians and other cultures of Southwest Asia associated Leo with the Sun, because the summer solstice occurred when the Sun was in that part of the sky.

Leo Minor is a recent addition to the constellations, introduced by Johannes Hevelius during the seventeenth century.

Gamma (γ) Leonis: This beautiful double star has orange-yellow components of 2nd and 3rd magnitude separated by 5 arc seconds.

R Leonis: This favorite Mira variable is easily found near Regulus. It ranges from magnitude 5.9 to 11 over about 10% months.

R Leonis Minoris: Another Mira star, taking about a year to vary between magnitudes 7.1 and 12.6.

M 65 and M 66: These two spiral galaxies near **Theta (θ) Leonis** are visible in

binoculars but give a better view in a telescope. Other interesting galaxies in Leo are **NGC 3628, M 95, M 96, M 105,** and **NGC 2903.**

Leonids: This meteor shower peaks annually on November 17. In 1966, observers recorded up to 30 meteors per second at its peak.

(Above, left) The Urania's Mirror (1825) is leaping rather than reclining. (Below) This view of Leo is dominated by blue-white Regulus and orange Delta (δ) Leonis in the sickle

The brighter members of the stunning Virgo Cluster are viewed in the view by M 86 (center) and M 84 (right), both elliptical galaxies. The majority of galaxies in the Cluster are 65 million light years from Earth

3C 273 Virginis: This is the brightest known quasar, but being only 13th magnitude, an 8 inch (200 mm) telescope is needed to identify it. Lying 3 billion light years away, it is the most distant object amateurs are likely to see in their telescopes.

naked eye symbol

telescope symbol

binoculars symbol

Many are within the range of binoculars or telescopes with front lens apertures of 2.4 inches (60 mm). Of course, most objects will reveal more detail when viewed with a larger telescope.

CONSTELLATION CHART FEATURES
Each of the constellation charts feature the following information (see sample pages, above):
• a pronunciation guide and the constellation's common name
• the genitive form of the constellation name, used when correctly naming objects within the constellation
• the standard three-letter abbreviation of the constellation's proper Latin name

• best viewing time, which is the approximate date the constellation is on meridian (that is, highest in the sky) at 10 pm at night—standard time, not daylight saving time
• a visibility grading, based on a scale from 1 to 4, representing the ease with which the constellation can be seen
• a hand symbol, which indicates the number of outstretched hand spans (each about 20 degrees across) that will cover the constellation east to west (left to right on the map). (See page 59 for an explanation of how to measure distances with your hand.)
• reference to specific sky chart(s), where the constellation is prominently featured
• where applicable, a symbol indicating that one of the 25

brightest stars is featured in the constellation (see 25 Brightest Stars table on p. 106)
• picture symbols indicating whether an object is first easily viewed with the unaided eye, binoculars, or a small telescope.

A FINAL WORD
The bimonthly sky charts and the constellation charts are easy to use, but take your time—especially on your first night out—and make sure you start with the correct sky chart. When focussing on one constellation, begin with an "easy" one, like Orion or Pegasus, or one of those with a visibility grading of 1, so you can gain a bit of confidence before taking on the more "elusive" constellations. Good luck, and dark skies!

The northern sky is full of bright stars now. With Auriga and Gemini almost overhead, the sky seems filled with bright stars like Capella, Castor, and Pollux. The Big Dipper is standing on its handle as it rises in the northeast. Leo is comfortably high in the east, its sickle appearing at its side.

High in the west, Perseus and Andromeda are well placed, and the Pleiades are also conspicuous. Often confused with the Little Dipper because of their shape, the

Pleiades in Taurus are a cluster of stars. The Little Dipper, hanging off Polaris, is far larger and harder to see.

WEST

CETUS

Mira

ERIDANUS

PISCES

30°

ECLIPTIC

ARIES

2ʰ
+30°

TAURUS

Aldebaran

60°

PEGASUS

Square of Pegasus

0ʰ

TRIANGULUM

Pleiades

4ʰ

ORION

Bellatrix

Betelgeuse

LACERTA

ANDROMEDA

PERSEUS

Capella

AURIGA

Zenith 10°N

Procyon

CYGNUS

Deneb

CEPHEUS

CASSIOPEIA

CAMELOPARDALIS

Zenith 50°N

Zenith 40°N

Zenith 30°N

GEMINI

Zenith 20°N

CANIS MINOR

Northern Cross

URSA MINOR

Polaris +90°

Castor

Pollux

6ʰ

9ʰ

NORTH

Small Dipper

URSA MAJOR

LYNX

8ʰ

HERCULES
Keystone

DRACO

Big Dipper

CANES VENATICI

LEO MINOR

CANCER

120°

Horizon 50°N

Horizon 40°N

Horizon 30°N

Horizon 20°N

Horizon 10°N

10ʰ

Sickle

SEXTANS

BOÖTES

COMA BERENICES

LEO

Regulus

13°

Arcturus

12ʰ

VIRGO

HYDRA

EAST

CRATER

18°

TIME*	
Jan 1	12 am
Jan 15	11 pm
Feb 1	10 pm
Feb 15	9 pm
etc.	

*Add one hour for DST

Wil Tirion

Orion is the chief attraction of the winter sky. High in the heavens, it is characterized by its three belt stars, lined up neatly, between brilliant blue-white Rigel and Betelgeuse. These nights, the sky has enough bright stars sprinkled liberally through it to form some interesting patterns. One of these is the "Heavenly G". Begin with Aldebaran, the bright red star in Taurus, and continue through the brightest stars of Gemini—Castor and Pollux to Procyon, in Canis Minor, and Canis Major's Sirius, the brightest star in the sky. The G's curve stops in Orion at Rigel, and then indents from Rigel to Betelgeuse.

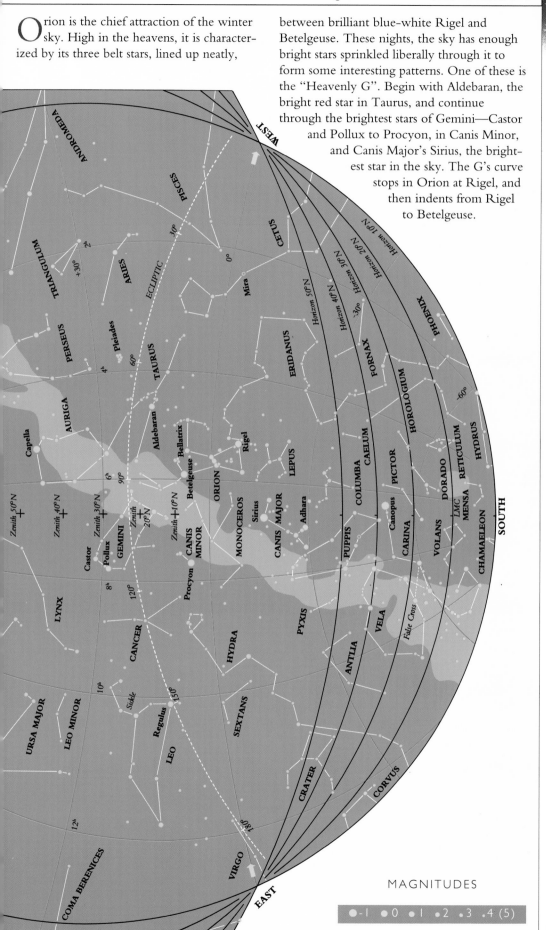

MAGNITUDES

●-1 ●0 ●1 •2 •3 •4 (5)

The Big Dipper is the centerpiece of the northern sky right now, and you can use it as a key to finding other stars. The two stars at the end of the bowl point toward Polaris, and the stars in the handle "arc to Arcturus", the center star in Boötes, the Herdsman. Looking much more like a kite than a herdsman, Boötes is the primary figure in the eastern sky.

The kite's northern neighbor is Corona Borealis, the Northern Crown—a semicircle of stars. Hercules is low in the east. In the western sky lie

Pollux and Castor, the bright twins of Gemini, with Capella in Auriga, the Charioteer, just to the north.

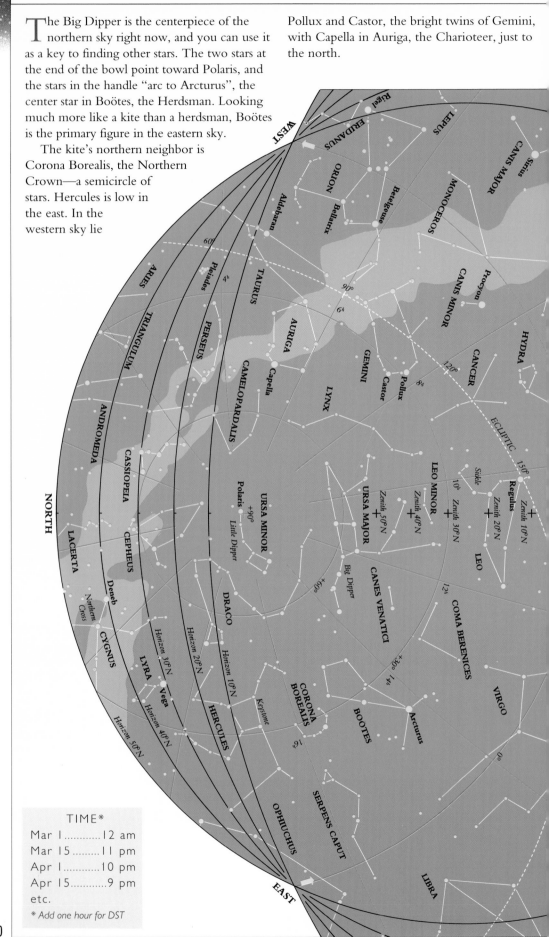

Riding high in the sky in both hemispheres, Leo's sickle has a distinctive shape like a backwards question mark. To find it, fill the Dipper's bowl with water and poke some holes in the bottom. As the water spills out, it will be Leo taking a shower!

Starting to set, the bright stars of Orion still dominate the western sky. The three stars shining in a row form his belt. In the east is a widely-spaced group of four bright stars called Virgo's Diamond. Spica, Arcturus, Denebola (the star at the eastern end of Leo) and Cor Caroli (the bright star in Canes Venatici) mark the corners of this fanciful stone.

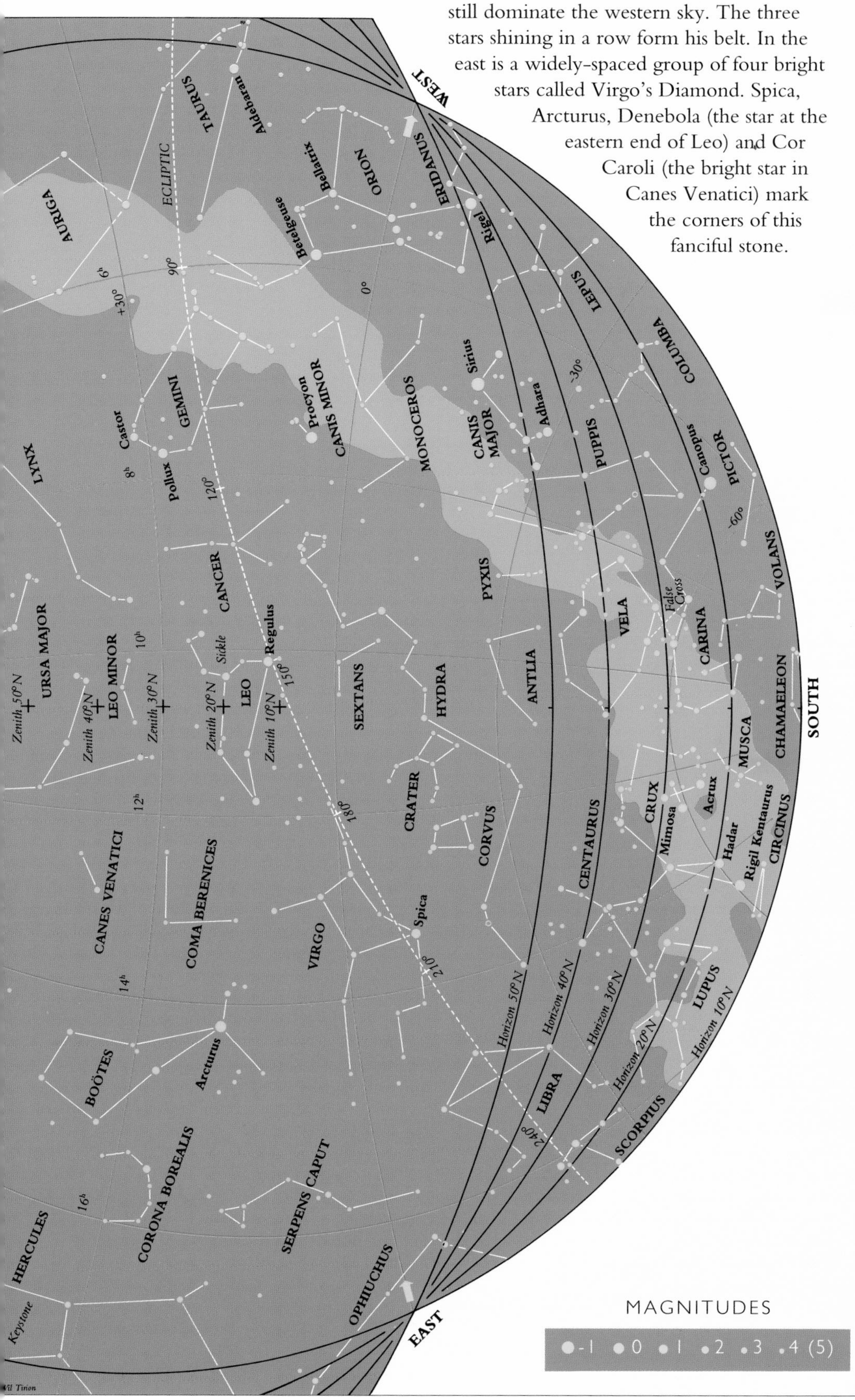

MAGNITUDES

-1 0 1 2 3 4 (5)

"Arc to Arcturus" is the chant for this chart. Although the Big Dipper is evident here, most other patterns are faint. Draco, the Dragon, takes up a good deal of room as it winds between the Dippers of Ursa Major and Ursa Minor, but its stars are difficult to see.

In the west, Leo's sickle is the salient feature. The eastern sky is much busier. In the northeast, the keystone of Hercules is easy to spot, as is the semicircle of stars that makes up Corona Borealis. The

Summer Triangle of Vega (in Lyra), Deneb (in Cygnus), and Altair (in Aquila) is rising, and the Milky Way is visible in a dark sky.

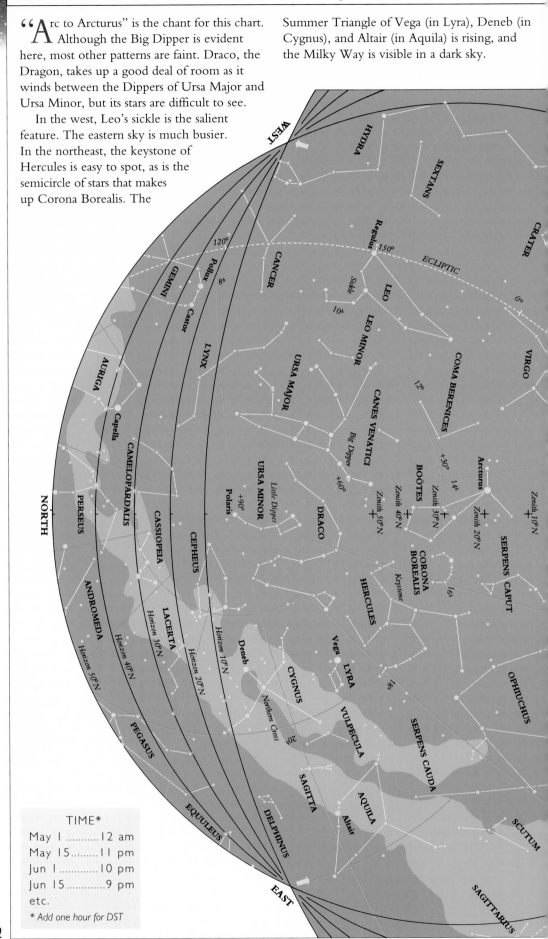

TIME*
May 1 .... 12 am
May 15 .... 11 pm
Jun 1 .... 10 pm
Jun 15 .... 9 pm
etc.
* Add one hour for DST

112

Although the Big Dipper is off the top of this chart, we can still use its handle to "arc to Arcturus". Now we can continue the line and "speed to Spica", the brightest star in the otherwise faint constellation of Virgo. Just to the southwest is the squarish shape of Corvus, the Crow, an often forgotten but striking constellation. The sickle of Leo is setting in the west.

The rich constellations of Scorpius and Sagittarius are just rising in the southeast. You are looking straight into the center of our galaxy here, and if the sky is dark enough you should be able to see the Milky Way.

WEST

CANCER

LYNX

URSA MINOR

ECLIPTIC

Regulus

150

Sickle

+30°

LEO MINOR

10ʰ

HYDRA

LEO

0°

SEXTANS

CRATER

Horizon 10° N

Horizon 20° N

Horizon 30° N

Horizon 40° N

Horizon 50° N

-30°

ANTLIA

VELA

Big Dipper

URSA MAJOR

CANES VENATICI

12ʰ

COMA BERENICES

CORVUS

-60°

Zenith 50° N

BOÖTES

14ʰ

Arcturus

VIRGO

Spica

HYDRA

CENTAURUS

Rigil Kentaurus

Hadar

Mimosa CRUX

Acrux

MUSCA

CARINA

Zenith 40° N

Zenith 30° N

Zenith 20° N

Zenith 10° N

210°

Vega

16ʰ

CORONA BOREALIS

SERPENS CAUPUT

LIBRA

240°

LUPUS

NORMA

CIRCINUS

ARA

TRIANGULUM AUSTRALE

APUS

PAVO

SOUTH

Keystone

HERCULES

OPHIUCHUS

Antares

SCORPIUS

Shaula

SAGITTARIUS

TELESCOPIUM

18ʰ

270°

LYRA

SERPENS CAUDA

SCUTUM

CORONA AUSTRALIS

Northern Cross

VULPECULA

Altair

AQUILA

300°

CAPRICORNUS

CYGNUS

20ʰ

SAGITTA

DELPHINUS

EAST

MAGNITUDES

●-1 ● 0 ● 1 • 2 • 3 • 4 (5)

Wil Tirion

Blueish-white Vega commands the center of the northern sky, with the other members of the Summer Triangle—Deneb and Altair—off to the east. The Milky Way is high enough for you to be able to make it out, unless your sky is badly light polluted. Off to the east, the mighty Square of Pegasus makes an appearance. The four stars are easy to see since there is not much else there. Farther eastward is the circlet of Pisces.

The western sky is full of strange shapes: the keystone of Hercules, the semicircle of Corona Borealis, and the kite of Boötes. The Big Dipper is low in the northwest.

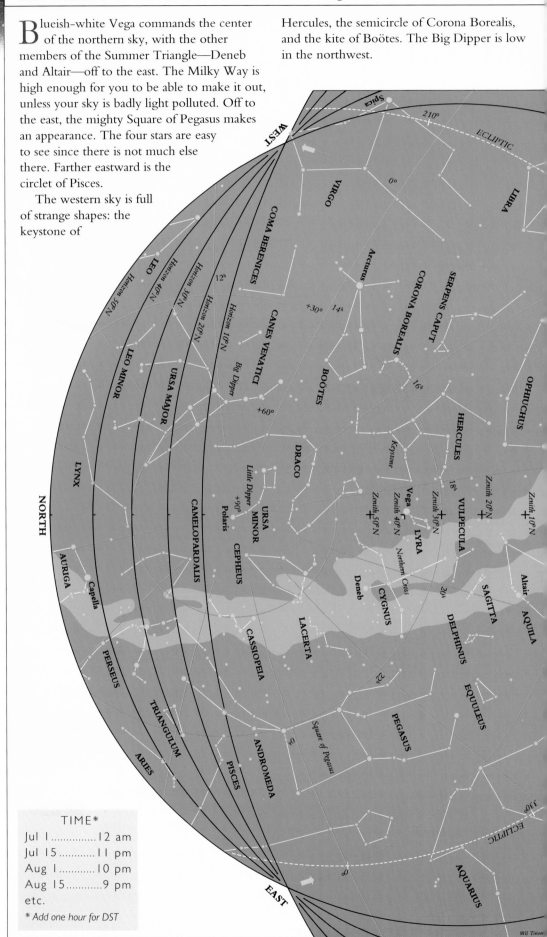

TIME*
Jul 1 .... 12 am
Jul 15 .... 11 pm
Aug 1 .... 10 pm
Aug 15 .... 9 pm
etc.
* Add one hour for DST

114

The Summer Triangle of Vega, Deneb, and Altair command this view of the sky. Although Ophiuchus, The Serpent Bearer, takes up a good deal of room in the southwest, most of its stars are faint and the constellation is difficult to decipher. Crossing Ophiuchus, Serpens snakes along toward the semicircle of Corona Borealis, with Arcturus and the kite of Boötes nearby.

The galactic center, with its constellations of Sagittarius and Scorpius, dominates the southern sky. Libra is just to the west. The large but faint figures of Capricornus, Aquarius, and the Square of Pegasus, rule the east.

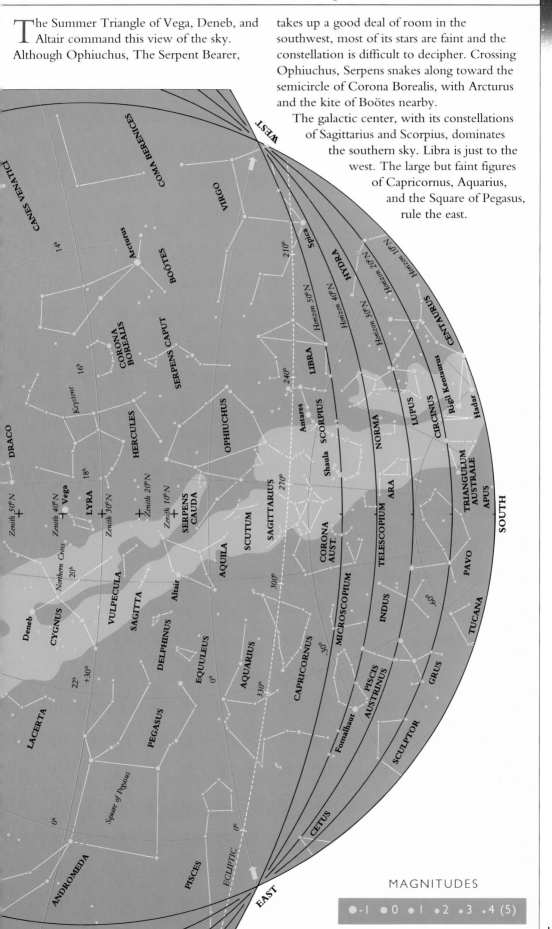

MAGNITUDES

-1   0   1   2   3   4 (5)

The misshapen W of Cassiopeia, the
Queen, is now high in the northeastern
sky. Her husband Cepheus, the King, looks
like an upside-down house with its roof
pointing roughly toward Polaris. To the east
lies daughter Andromeda, still chained to a
rock. Her savior, Perseus, is to the northeast,
and Pegasus, his winged horse, is to
Andromeda's south. There is also
a giraffe (Camelopardalis) in
the picture, but its stars
are so faint that you
can easily miss this
constellation.

In ancient Mediterranean lore, whenever
Capella rose in the evening, as on this chart,
winter storms would not be far behind.

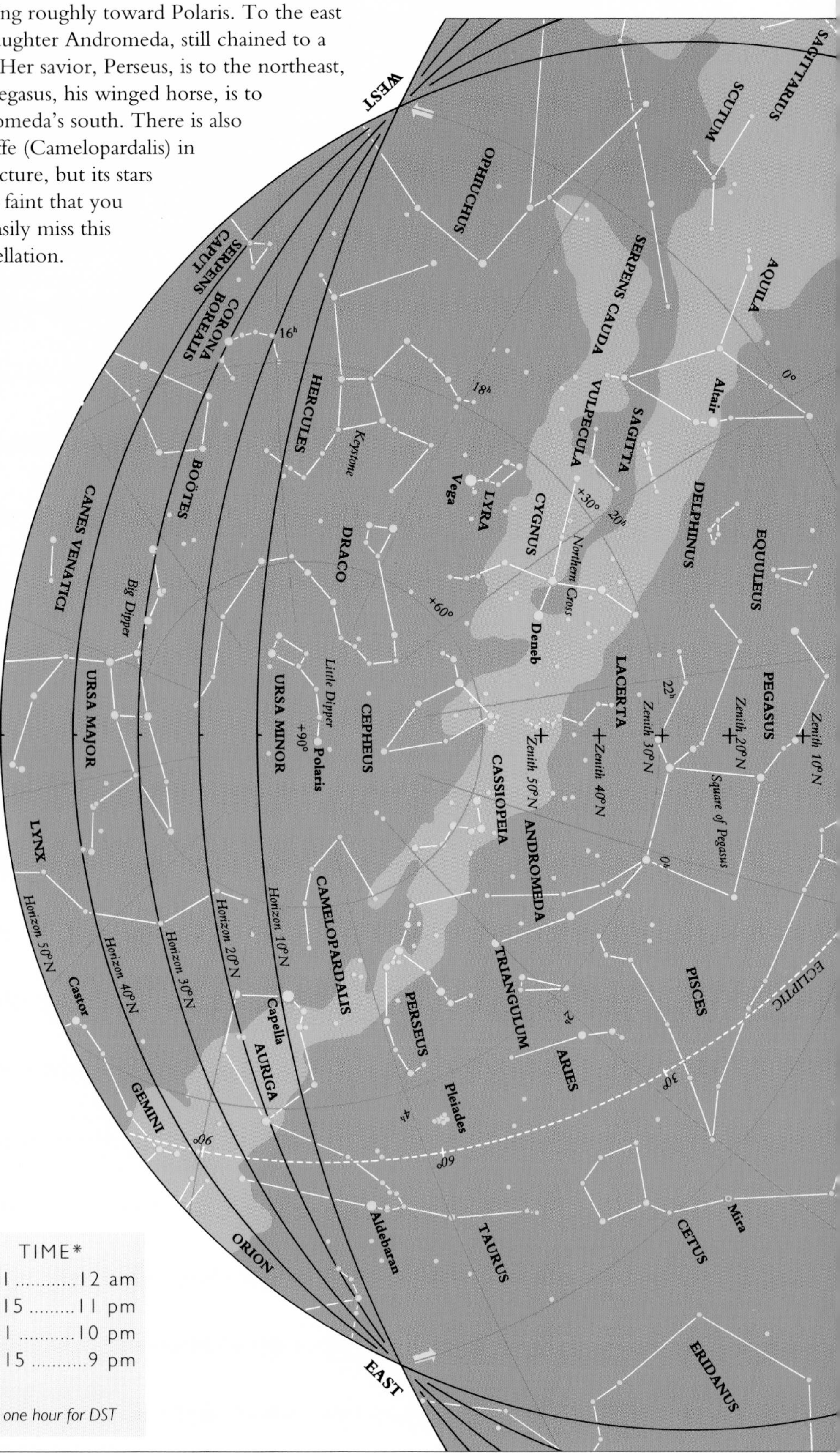

TIME*
Sep 1 ....12 am
Sep 15 ....11 pm
Oct 1 ....10 pm
Oct 15 ....9 pm
etc.
* Add one hour for DST

With the Summer Triangle setting in the west, the Great Square of Pegasus now governs the sky. The distinctive circlet of faint stars forming the head of Pisces is immediately to the south. Far to the south of the Square of Pegasus, past Aquarius, is lonely Fomalhaut—the bright star in Piscis Austrinus.

The eastern sky, with few bright stars and few easy patterns, is dominated by the aqueous constellations of Pisces, the Fish; Cetus, the Whale; and Eridanus, the River.

Aldebaran, the bloodshot red eye of Taurus, the Bull, is just rising, and the compact structure of the Pleiades lies above it in the northeast.

WEST

Keystone

HERCULES

OPHIUCHUS

Vega

LYRA

18°

SERPENS CAUDA

SCUTUM

SAGITTARIUS

270°

Scutum

SCORPIUS

Deneb

Northern Cross

20°

CYGNUS

VULPECULA

SAGITTA

DELPHINUS

Altair

AQUILA

CAPRICORNUS

300°

CORONA AUST

ARA

MICROSCOPIUM

TELESCOPIUM

PAVO

EQUULEUS

330°

PISCIS AUSTRINUS

Fomalhaut

INDUS

GRUS

OCTANS

SOUTH

22ʰ

PEGASUS

Square of Pegasus

Zenith 20°N

Zenith 30°N

0ʰ

Zenith 40°N

Zenith 50°N

LACERTA

ANDROMEDA

Zenith 10°N

0ʰ

AQUARIUS

CETUS

SCULPTOR

Horizon 50°N

Horizon 40°N

PHOENIX

TUCANA

SMC

HYDRUS

Achernar

60°

HOROLOGIUM

2ʰ

PISCES

30°

ECLIPTIC

TRIANGULUM

ARIES

Mira

0°

ERIDANUS

FORNAX

Horizon 30°N

Horizon 20°N

Horizon 10°N

+30°

PERSEUS

4ʰ

Pleiades

60°

TAURUS

Aldebaran

ORION

EAST

AURIGA

MAGNITUDES

● -1 ● 0 ● 1 • 2 • 3 • 4 (5)

Wil Tirion

The eastern sky is now rich with bright stars, ranging from Capella to Castor and Pollux (the twins) and the dog stars Procyon and Sirius. If your sky is dark enough, you will be able to see the Milky Way arching overhead. It is rich in the west, as it goes through Cygnus, but thins out in the east. Far from the galactic center, this is the Milky Way's dimmest part.

Cassiopeia straddles the top of the sky, its W inverted into an M. Since the pointer stars of the Big Dipper may be

too low at the moment to help find the pole, we can use Cepheus, whose pointed roof suggests the way toward Polaris.

WEST

NORTH

EAST

AQUARIUS

DELPHINUS

EQUULEUS

SAGITTA

VULPECULA

LYRA

Vega

Keystone

HERCULES

BOÖTES

CANES
VENATICI

CYGNUS

Deneb

Northern Cross

CEPHEUS

DRACO

Little Dipper

URSA MINOR

Polaris
+90°

Big Dipper

LEO
MINOR

URSA MAJOR

LEO

Sickle

CAMELOPARDALIS

CASSIOPEIA

LACERTA

PEGASUS

Square of Pegasus

+30°

ANDROMEDA

PERSEUS

Capella

LYNX

AURIGA

Castor

Pollux

CANCER

HYDRA

GEMINI

Procyon
CANIS MINOR

MONOCEROS

Sirius
CANIS MAJOR

ORION

Bellatrix

Betelgeuse

Aldebaran

TAURUS

Pleiades

Zenith 40°N

TRIANGULUM

Zenith 30°N

ARIES

Zenith 20°N

PISCES

Zenith 10°N

CETUS

ECLIPTIC

Horizon 50°N

Horizon 40°N

Horizon 30°N

Horizon 20°N

Horizon 10°N

 Zenith 50°N

+60°

33h

20h

22h

0h

2h

4h

6h

8h

10h

12h

0°

30°

60°

118

The sky to the south seems divided into two camps. The Square of Pegasus overlooks the west. To the south is Aquarius, and Fomalhaut in Piscis Austrinus. Orion rules the eastern camp, its three belt stars pointing westward toward Aldebaran in Taurus, and eastward to Sirius.

A watery border between east and west is formed by Eridanus, the River, and Cetus, the Whale. Two small constellations continue this border. One is Aries, the Ram, and the other, consisting of three stars, is somewhat unimaginatively called Triangulum.

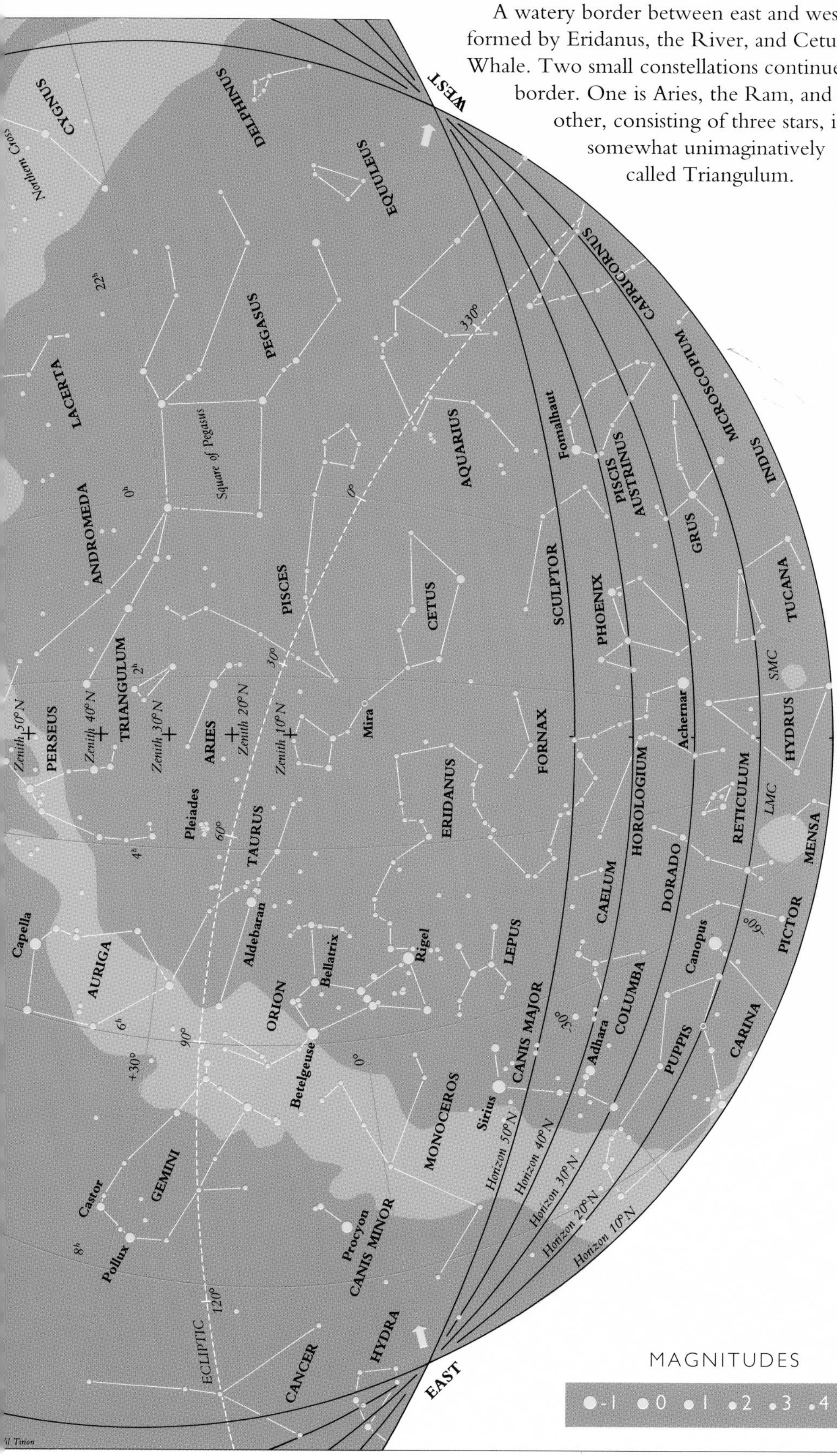

MAGNITUDES

●-1 ●0 ●1 •2 •3 •4 (5)

Il Tirion

119

SKY CHART 7 Southern Latitudes Looking North

Like the Dipper in the north, Orion is a useful key for pointing to other groups of stars in the sky. Draw a line from Rigel through Betelgeuse to find Gemini, the Twins. Then you can go southeast from the three belt stars to Sirius, the brightest star in both hemispheres, or northwest to Aldebaran in Taurus.

Just east of Orion is Monoceros, the Unicorn. The animal might be one of the most wonderful creatures ever created by human imagination, but this version is full of faint stars and is undistinguished. Off to the east is the huge expanse of Hydra, the Serpent, and off to the west is Eridanus, the River, stretching almost as far.

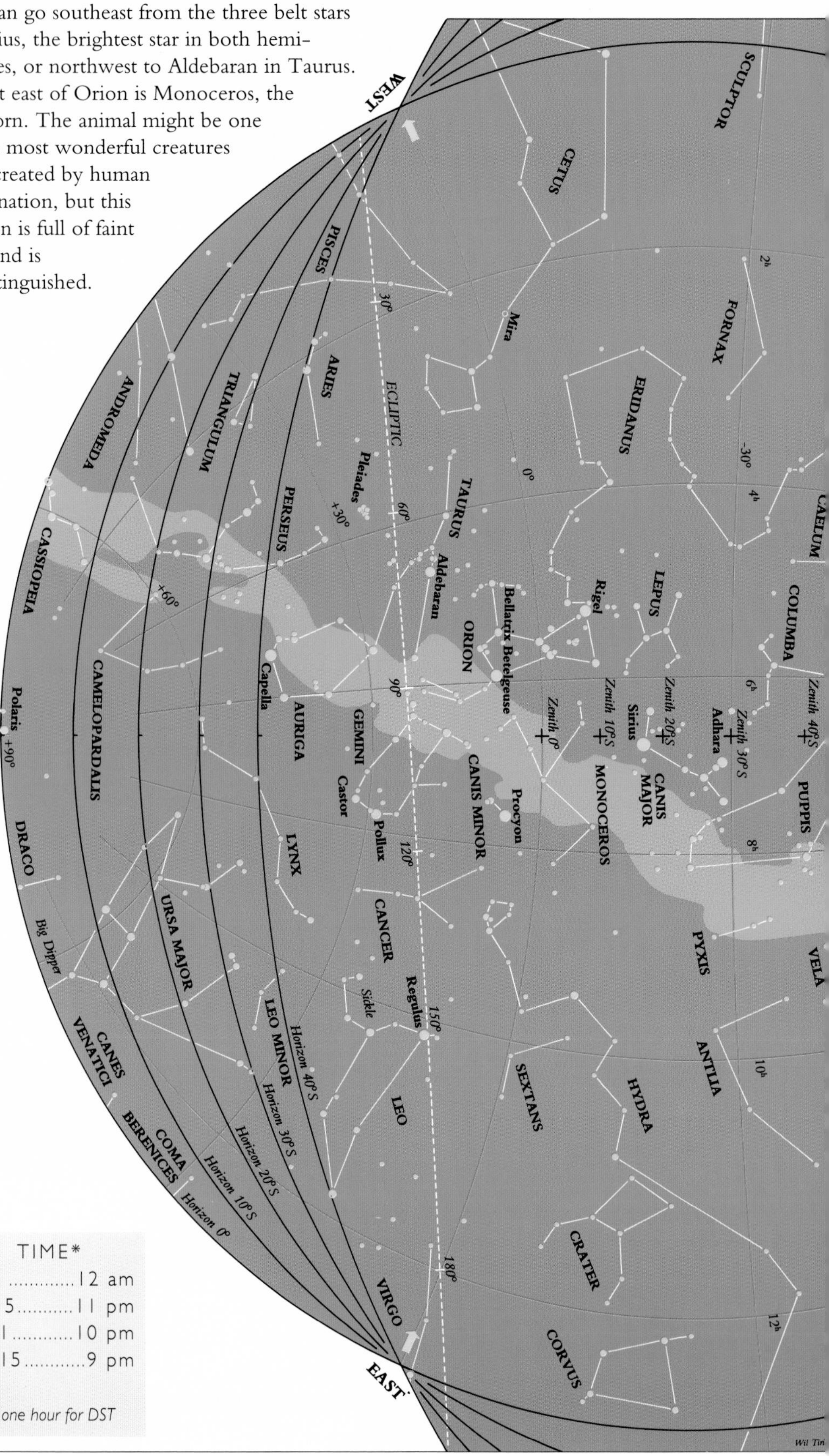

TIME*

Jan 1 .... 12 am
Jan 15.......... 11 pm
Feb 1 .... 10 pm
Feb 15 .... 9 pm
etc.

* *Add one hour for DST*

S irius and Canopus, the brightest and second
brightest stars in the sky, command our
view tonight, with Rigel and Achernar
completing a large semicircle overhead. The
sky is a narrow oasis of beauty, with the Milky
Way straddling such lovely constellations as
Puppis, Vela, Carina, and Crux. Just to
the west lies the faint glow of the Large
Magellanic Cloud (LMC), now at the crest
of its tight circuit around the celestial pole.
Deep in the southeast lies the stunning
constellation of Centaurus,
surrounding on three sides the
tiny constellation of Crux,
the Southern Cross.

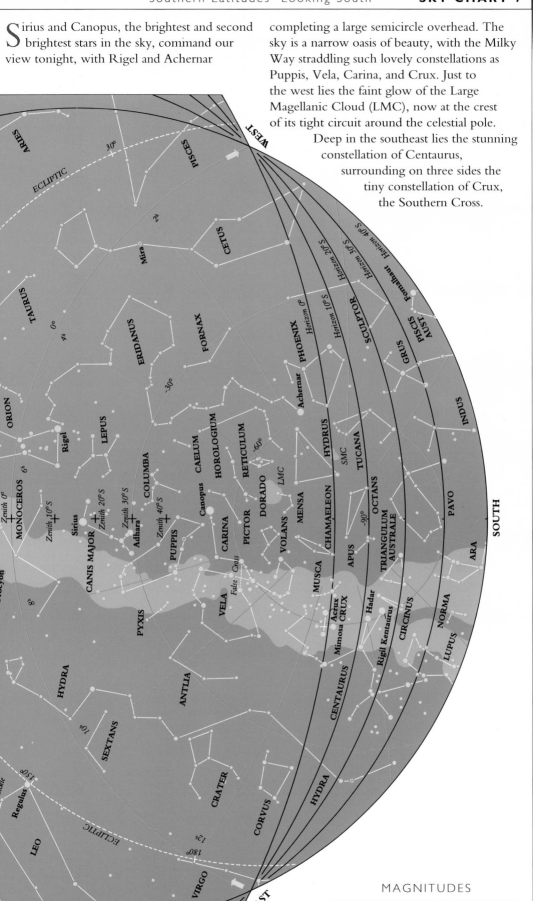

MAGNITUDES

●-1 ●0 ●1 •2 •3 •4 (5)

Regulus in Leo marks the sky's center at this hour, and a strange sight the sky is for visitors from the Northern Hemisphere. Leo's sickle, like everything else, appears upside down! For Southern Hemisphere dwellers, this is how Leo normally appears, just as Orion, in the west, has Rigel at the top and Betelgeuse at the bottom.

Spica and Arcturus outshine all other stars in the eastern sky, although their constellations of Virgo and Boötes are otherwise unremarkable.

Crater and Corvus are high in the east. With its brighter stars and rhomboid appearance, Corvus is the easier of the two to spot.

WEST

NORTH

EAST

ERIDANUS
Rigel
Bellatrix
ORION
Betelgeuse
TAURUS
AURIGA
Capella
Horizon 0°S
CAMELOPARDALIS
Horizon 10°S
Horizon 20°S
Horizon 30°S
Horizon 40°S
GEMINI
Castor
Pollux
LYNX
URSA MINOR
Little Dipper
+90°
URSA MAJOR
Big Dipper
DRACO
LEO MINOR
CANES VENATICI
HERCULES
CORONA BOREALIS
BOÖTES
Arcturus
COMA BERENICES
SERPENS CAPUT
OPHIUCHUS
VIRGO
Spica
CANCER
LEO
Sickle
Regulus
+30°
+60°
90°
120°
150°
180°
210°
240°
CANIS MINOR
Procyon
MONOCEROS
LEPUS
CANIS MAJOR
Sirius
Adhara
COLUMBA
6ʰ
PUPPIS
8ʰ
-30°
0°
HYDRA
SEXTANS
CRATER
CORVUS
HYDRA
PYXIS
ANTLIA
VELA
10ʰ
12ʰ
CENTAURUS
LUPUS
LIBRA
ECLIPTIC
14ʰ
16ʰ
Antares
SCORPIUS
Zenith 0°
Zenith 10°S
Zenith 20°S
Zenith 30°S
Zenith 40°S

TIME*
Mar 1 ....12 am
Mar 15 ....11 pm
Apr 1 ....10 pm
Apr 15 ....9 pm
etc.
* Add one hour for DST

The southern Milky Way is here marching across the sky in all its glory. An elegant paradigm of stars begins with Alpha (α) Centauri (or Rigil Kentaurus) and Beta (β) Centauri, the "pointers" to the Southern Cross. The parade goes on through the rich star clouds of Carina and the bright star Canopus, ending with Sirius. There is little in the northern or southern sky to match this.

In the south are the faint constellations around the pole, to which the Cross is a pointer. Octans, Tucana, Reticulum, and Phoenix are something of a wasteland, but if your sky is dark you will see the two Magellanic Clouds.

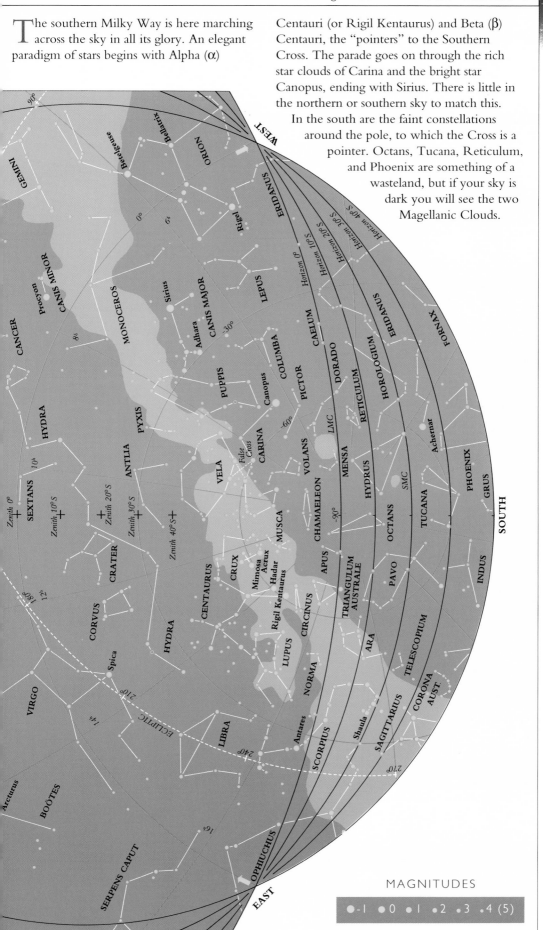

MAGNITUDES

●-1 ● 0 ● 1 ● 2 ● 3 ● 4 (5)

Vit Tirion

With Arcturus in the center, the sky at this point is filled with a variety of constellations. The galactic center is high in the southeastern heavens, its majestic forms of Sagittarius and Scorpius adding sparkle to a dark night. Rising in the northeast are Lyra, the Lyre, its bright star Vega leading the way, and Aquila, the Eagle, marked by Altair, its brightest star. Ophiuchus is a centerpiece of the eastern sky, its wide area making up for its lack of bright stars.

Leo, Virgo, Corvus, and Crater

dominate the west, this wide area being punctuated by only two bright stars—Regulus (in Leo) and Spica (in Virgo).

TIME*	
May 1	12 am
May 15	11 pm
Jun 1	10 pm
Jun 15	9 pm
etc.	

* Add one hour for DST

From a dark sky, you can see the Milky Way in all its splendor stretch across the sky, through Aquila and Scutum to the galactic center in Sagittarius, and on to Scorpius. The Milky Way continues through Centaurus, with the "pointers" Alpha (α) Centauri (or Rigil Kentaurus) and Beta (β) Centauri pinpointing Crux, now high in the south, and Vela, and Puppis. If our view of the center were not blocked by dark matter, the whole sky would appear much brighter. Farther south shine the Large and Small Magellanic Clouds, our two neighboring galaxies.

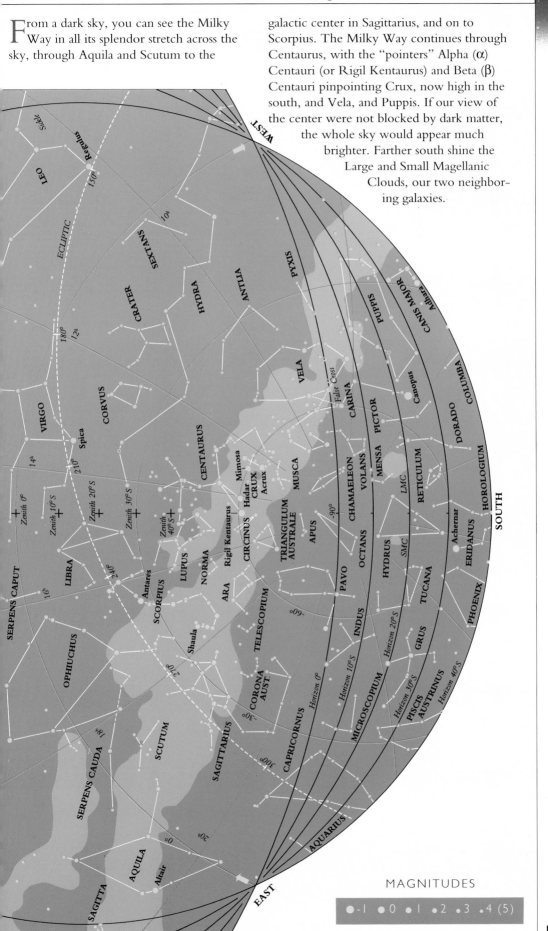

MAGNITUDES

-1 0 1 2 3 4 (5)

Two birds fly through the northern sky here: Cygnus, the Swan, and Aquila, the Eagle, with its bright star Altair. Nearby, bright star Vega is crossing the meridian. Arching directly overhead, the Milky Way, with its bright heart in Sagittarius, should be visible, unless you are in the city. The constellations Delphinus, the Dolphin, and Sagitta, the Arrow, are small but distinct.

To the east, Capricornus and Aquarius are large but faint, while Piscis Austrinus, the Southern Fish, is easy to find only because of Fomalhaut, its one bright star.

WEST

HYDRA

Spica

VIRGO

ECLIPTIC

210°

-30°

14ʰ

CENTAURUS

LUPUS

16ʰ

LIBRA

COMA BERENICES

Arcturus

BOÖTES

CORONA BOREALIS

0°

+30°

SERPENS CAPUT

240°

Antares

Shaula

SCORPIUS

CANES VENATICI

URSA MAJOR

Big Dipper

+60°

Keystone

HERCULES

OPHIUCHUS

270°

Zenith 40°S

18ʰ

CORONA AUST.

DRACO

Little Dipper

LYRA

Vega

SERPENS CAUDA

Zenith 0°

Zenith 10°S

Zenith 20°S

Zenith 30°S

SCUTUM

SAGITTARIUS

URSA MINOR

+90°

CEPHEUS

CYGNUS

Deneb

Northern Cross

SAGITTA

Altair

AQUILA

300°

20ʰ

MICROSCOPIUM

GRUS

NORTH

CASSIOPEIA

Horizon 0°

Horizon 10°S

Horizon 20°S

LACERTA

Horizon 30°S

Horizon 40°S

VULPECULA

DELPHINUS

EQUULEUS

CAPRICORNUS

330°

22ʰ

Square of Pegasus

PEGASUS

PISCES AUSTRINUS

Fomalhaut

ANDROMEDA

PISCES

AQUARIUS

0ʰ

CETUS

SCULPTOR

EAST

TIME*
Jul 1 ....12 am
Jul 15............11 pm
Aug 1 ....10 pm
Aug 15 ....9 pm
etc.
* Add one hour for DST

126

The sky is dominated by the glowing center of the Milky Way in Sagittarius, directly overhead. From there, the river of stars flows through the tail of Scorpius down to the Southern Cross and Carina.

Lying mainly between the bright stars Fomalhaut and Achernar in the southeast, and the Milky Way, you can find several birds, namely Grus, the Crane; Pavo, the Peacock; Tucana, the Toucan; and the fabulous Phoenix. Apus, the Bird of Paradise, is closer to the Southern Cross.

These birds compete with Musca, the Fly, and Volans, the Flying Fish.

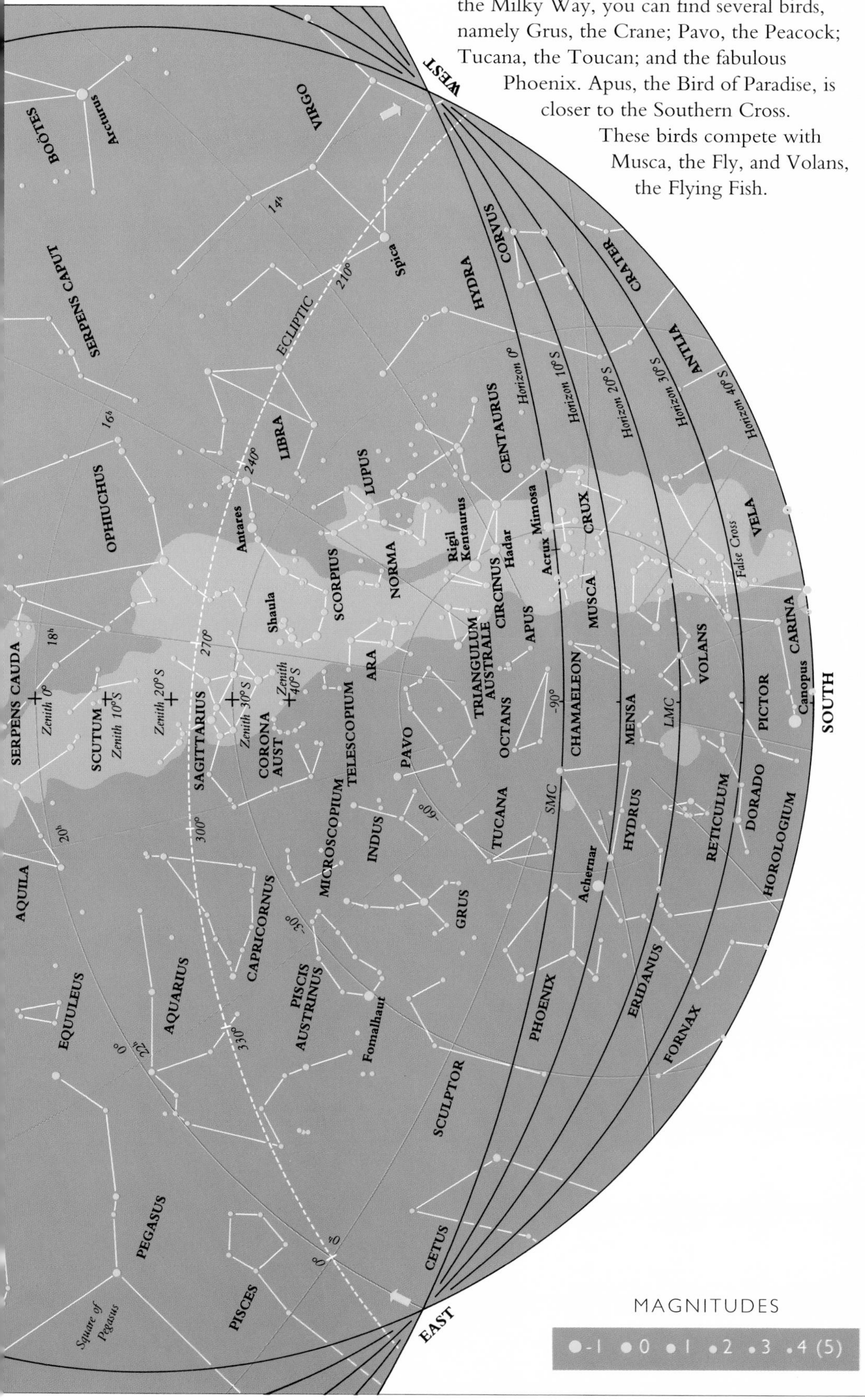

MAGNITUDES

-1 0 1 2 3 4 (5)

Dominated by Fomalhaut, the bright star now crossing the meridian, this view of the sky is unusual in that the entire eastern half lacks a single major star. The large but faint constellations of Eridanus, Cetus, and Pisces overlook this wintry portion of the sky, but it is the Square of Pegasus that really dominates.

The constellations of the Milky Way enrich the north and west. As you look farther west, the Milky Way widens as it approaches the

galactic center in Sagittarius. Capricornus, Microscopium, and Corona Australis complete this panorama.

WEST

ECLIPTIC

SCORPIUS

Shaula

18ʰ

OPHIUCHUS

270°

CORONA AUSTRALIS

20ʰ

HERCULES

SERPENS CAUDA

SAGITTARIUS

SCUTUM

-30°

300°

Keystone

LYRA

Vega

+30°

VULPECULA

SAGITTA

Altair

AQUILA

0°

MICROSCOPIUM

DRACO

Northern Cross

DELPHINUS

EQUULEUS

CAPRICORNUS 330°

PISCIS AUSTRINUS

22ʰ

GRUS

CYGNUS

Deneb

+60°

AQUARIUS

PEGASUS

Zenith 0°

Zenith 10°S

Zenith 20°S

Zenith 30°S

Fomalhaut

Zenith 40°S

PHOENIX

0ʰ

URSA MINOR

Small Dipper

+90° Polaris

CEPHEUS

LACERTA

Square of Pegasus

0°

SCULPTOR

NORTH

CASSIOPEIA

ANDROMEDA

PISCES

CETUS

2ʰ

CAMELOPARDALIS

PERSEUS

TRIANGULUM

Horizon 30°S

ARIES

30°

Mira

FORNAX

Horizon 40°S

Horizon 20°S

Pleiades

Horizon 10°S

Horizon 0°

Aldebaran

TAURUS

ERIDANUS

4ʰ

EAST

TIME*
Sep 1.............12 am
Sep 15..........11 pm
Oct 1............10 pm
Oct 15............9 pm
etc.
* Add one hour for DST

A string of bright stars running southward livens up this view of the southern sky. Fomalhaut in Piscis Austrinus, the Southern Fish, is the northernmost star. The line runs through Phoenix to Achernar in Eridanus, the River, and ends in Canopus, the brightest star in Carina. The western sky is full of rich Milky Way constellations, with the center of the galaxy high in the west.

The eastern half of this sky is remarkably devoid of bright stars, with Cetus and Eridanus taking up most of the space. The Magellanic Clouds are again rising to prominence in the south.

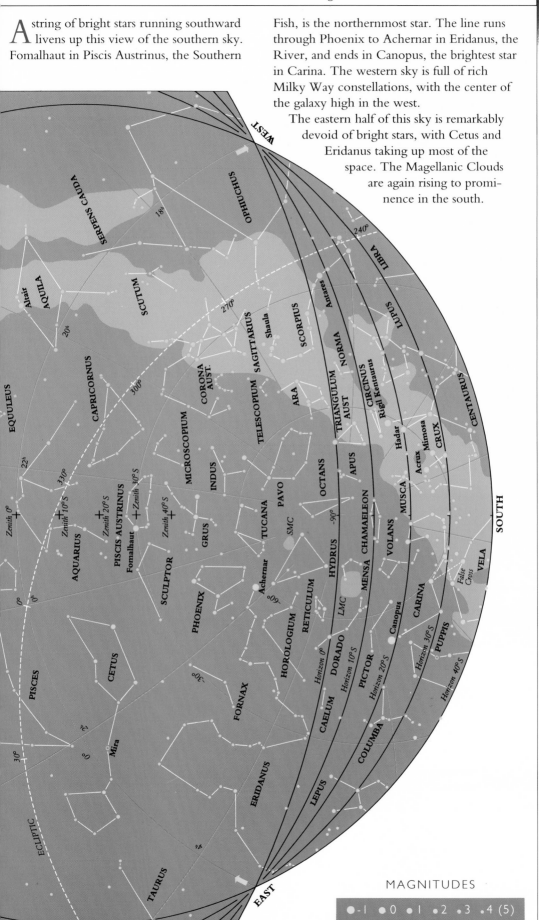

MAGNITUDES

● -1 ● 0 ● 1 ● 2 · 3 · 4 (5)

Back in the sky once more, Orion outshines all the other constellations in this view. To the east are Sirius and Canis Major. Lepus, the Hare, scampers to the south, under the watchful eye of Columba, the Dove. The bright star Aldebaran is in the north, with Capella closer to the horizon. Perseus is crossing the meridian, low in the north, with Cetus, the Whale; Eridanus, the River; and Fornax, the Furnace, higher in the sky.

With its large but faint constellations like Pegasus, Pisces, and Cetus, the west forms quite a contrast, Fomalhaut being the only bright star in the area.

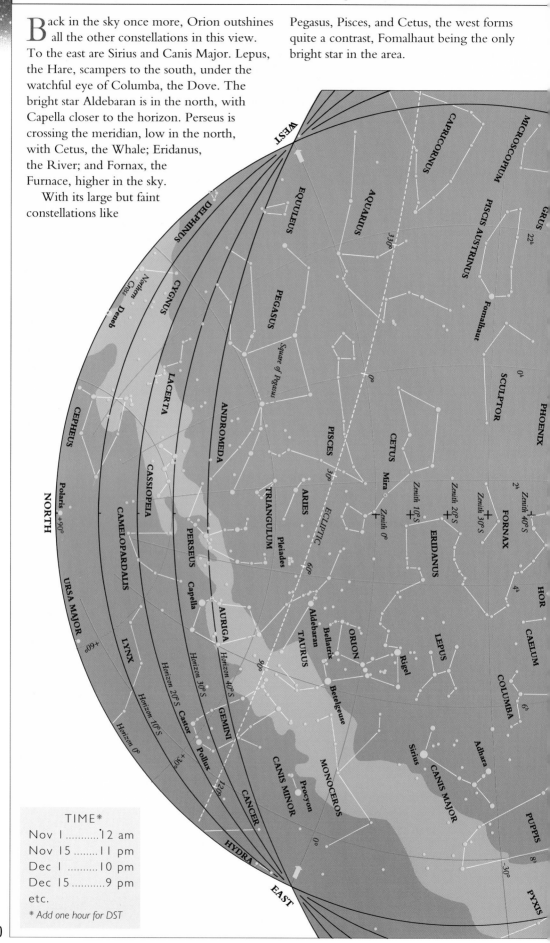

TIME*
Nov 1 .... 12 am
Nov 15 .... 11 pm
Dec 1 .... 10 pm
Dec 15 .... 9 pm
etc.
* Add one hour for DST

Four stars dominate this view: Betelgeuse, Rigel, Sirius, and Canopus. A fifth, Procyon, is just rising in the east. In the southeast are parts of a great dismembered ship—once a vast constellation known as Argo Navis—that must have given pause to south sea explorers: Puppis, the Stern; Carina, the Keel; Vela, the Sail; and Pyxis, the Compass.

There are also birds of all shades flying about: Columba, the Dove, and to the west, Tucana, Pavo, Apus, Grus, and Phoenix. Elsewhere are several small constellations, including Microscopium, Indus, Telescopium, Ara, and Triangulum Australe.

WEST

PEGASUS

EQUULEUS

AQUARIUS

Square of Pegasus

PISCES

CAPRICORNUS

300°

330°

ECLIPTIC

22h

SAGITTARIUS

CORONA AUSTRALIS

PISCIS AUSTRINUS

Fomalhaut

MICROSCOPIUM

INDUS

PAVO

TELESCOPIUM

0h

SCULPTOR

GRUS

SCORPIUS

CETUS

PHOENIX

TUCANA

HYDRUS

OCTANS

ARA

NORMA

PISCES

2h

CETUS

FORNAX

Achernar

RETICULUM

SMC

MENSA

CHAMAELEON

APUS

TRIANGULUM AUSTRALE

CIRCINUS

Rigil Kentaurus

LUPUS

Mira

Zenith 0° S

Zenith 10° S

Zenith 20° S

Zenith 30° S

Zenith 40° S

ERIDANUS

HOROLOGIUM

LMC

VOLANS

CARINA

MUSCA

Acrux

Mimosa

Hadar

CRUX

-90°

SOUTH

4h

DORADO

PICTOR

CAELUM

COLUMBA

Canopus

-60°

VELA

CENTAURUS

LEPUS

Rigel

Bellatrix

Adhara

CANIS MAJOR

PUPPIS

-30°

Fake Cross

ORION

Betelgeuse

6h

0°

Sirius

Horizon 0°

PYXIS

Horizon 10° S

ANTLIA

Horizon 20° S

Horizon 30° S

Horizon 40° S

GEMINI

MONOCEROS

8h

Procyon

CANIS MINOR

CANCER

EAST

ECLIPTIC

Tirion

MAGNITUDES

●-1 ● 0 ● 1 • 2 • 3 • 4 (5)

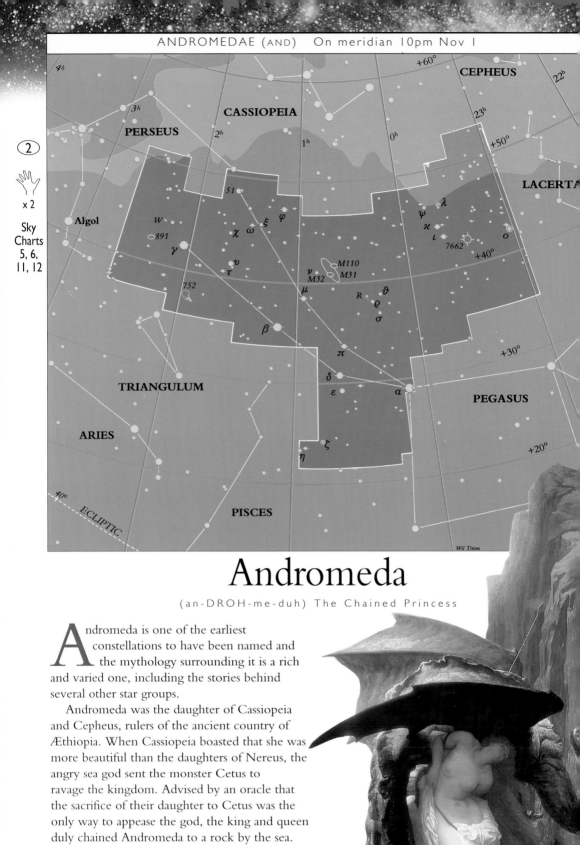

CEPHEUS

CASSIOPEIA

PERSEUS

LACERTA

Algol

Sky
Charts
5, 6,
11, 12

x 2

891

7662

M110
M32   M31

TRIANGULUM

PEGASUS

ARIES

PISCES

ECLIPTIC

Wil Tirion

Andromeda

(an-DROH-me-duh) The Chained Princess

Andromeda is one of the earliest
constellations to have been named and
the mythology surrounding it is a rich
and varied one, including the stories behind
several other star groups.

Andromeda was the daughter of Cassiopeia
and Cepheus, rulers of the ancient country of
Æthiopia. When Cassiopeia boasted that she was
more beautiful than the daughters of Nereus, the
angry sea god sent the monster Cetus to
ravage the kingdom. Advised by an oracle that
the sacrifice of their daughter to Cetus was the
only way to appease the god, the king and queen
duly chained Andromeda to a rock by the sea.
Perseus came to the rescue just in time, however,
swooping down astride the winged horse
Pegasus. Perseus was able to save Andromeda
from her cruel fate by revealing the hideous head
of Medusa to Cetus, which instantly turned the
great monster to stone.

*A detail from a painting by Frederic Leighton (1830–96),
showing Cetus about to devour the helpless Andromeda*

...he bright stars of Andromeda, featuring the Andromeda ...alaxy—the nearest major galaxy to Earth

Although Andromeda is justly famous for the great and distant galaxy that resides within the constellation, its stars are not very bright. It is easy to find, however, located south of Cassiopeia's W, and just off one corner of the Great Square of Pegasus. In fact, Alpheratz, the star at the northeastern corner of the square of Pegasus, belongs to Andromeda.

The Andromeda Galaxy (M 31): The closest major galaxy to us, the Andromeda Galaxy was first thought to be a nebula, and was listed in comet hunter Charles Messier's eighteenth-century catalogue of nebulae. A spiral galaxy much like our own Milky Way, it is a maelstrom comprising 200 billion suns and clouds of dust and gas. It is bright enough to be seen with binoculars from city sites and with the naked eye beneath a dark sky, being one of the most distant objects visible to the unaided eye. In the field of larger binoculars, or using a small telescope, you can see its two neighboring elliptical galaxies. **M 32** is small and compact; **M 110** is larger and more diffuse, and is therefore harder to see.

Gamma (γ) Andromedae: This is a beautiful double star. The brighter member of the pair is a golden yellow, and its companion is greenish blue.

R Andromedae: This Mira star has a range of 9 magnitudes.

The Andromeda Galaxy, with M 32 (left) and M 110 (right). Only the central yellow regions appear in a small telescope.

NGC 752: This open cluster lies about 5 degrees south of Gamma (γ) Andromedae and is easy to find because of its relatively bright stars. Because it is spread out over such a large area, it is actually easier to see through binoculars than through a telescope. If using a telescope, use it at its lowest power.

NGC 7662: A fairly bright planetary nebula, this blue-green object looks almost starlike through the smallest telescopes. But through a 6 inch (150 mm) telescope at moderate power, it becomes a graceful, glowing spot of gas about 30 arc seconds across.

NGC 891: This galaxy is a challenge even for 6 inch (150 mm) telescopes. However, with good eyes and a dark sky, you will see one of the best examples of a spiral galaxy, viewed edge-on.

Andromeda! Sweet woman!

why delaying, So timidly

among the stars: come hither!

Join this bright throng, and

nimbly follow whither

They all are going.

Endymion, JOHN KEATS (1795-1821), English poet

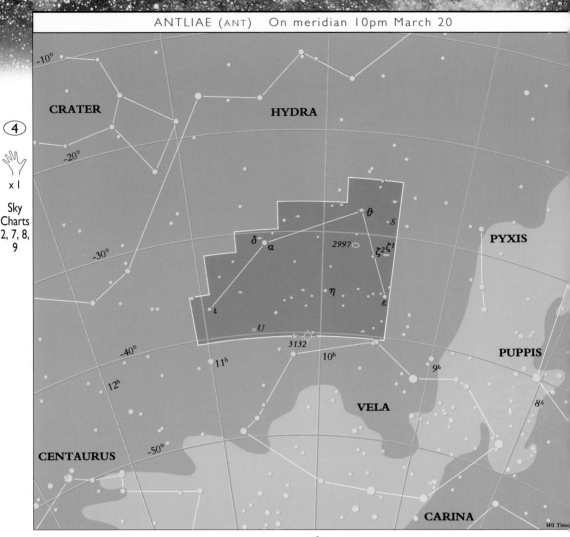

-10°

CRATER   HYDRA

-20°

4

✋
x 1

Sky
Charts
2, 7, 8,
9

-30°

PYXIS

ϑ

S

δ

α

2997

ζ² ζ¹

η

ε

ι

U

-40°

3132

11ʰ

10ʰ

9ʰ

12ʰ

PUPPIS

8ʰ

VELA

CENTAURUS   -50°

-50°

CARINA

Wil Tirion

Antlia

(ANT-lee-uh) The Air Pump

Antlia Pneumatica, the Air Pump, named after seventeenth-century physicist Robert Boyle's invention, is a southern constellation. It was given this somewhat unpoetic name by Nicolas-Louis de Lacaille during the time he spent working at an observatory at the Cape of Good Hope, from 1750 to 1754. As a result of his observations of some 10,000 southern stars, de Lacaille divided the far southern sky into 14 new constellations, of which Antlia is one.

Antlia is a small, faint constellation just off the bright southern Milky Way, not far from Vela and Puppis. Its alpha (α) star is just barely the constellation's brightest star and has been given no proper name. It is quite red in color and possibly varies slightly in magnitude.

🔭 **NGC 2997**: This is a large, faint spiral galaxy, with a stellar nucleus. It is quite difficult to observe with a small telescope.

NGC 2997. An impressive galaxy with its spiral arms traced out by blue stars, pink hydrogen clouds and dust

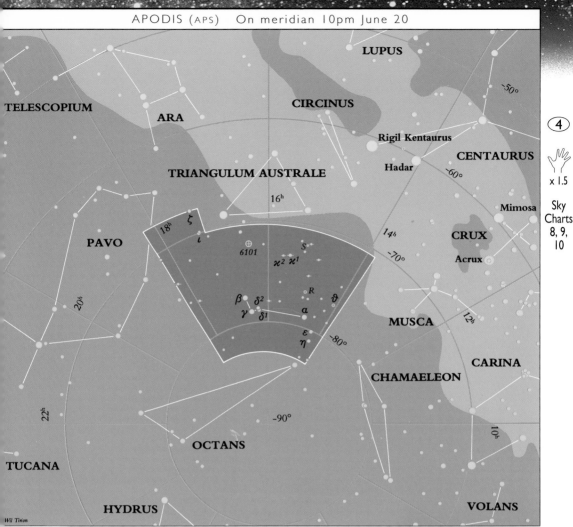

LUPUS

TELESCOPIUM

CIRCINUS

ARA

Rigil Kentaurus

CENTAURUS

-50°

-60°

TRIANGULUM AUSTRALE

Hadar

16ʰ

Mimosa

18ʰ ζ

ι

⊕
6101

14ʰ

CRUX

PAVO

S

π² π¹

-70°

Acrux

20ʰ

β δ²
γ δ¹
a

R

ϑ

12ʰ

MUSCA

ε
η

-80°

CARINA

CHAMAELEON

22ʰ

-90°

10ʰ

OCTANS

TUCANA

VOLANS

HYDRUS

Wil Tirion

④

🖐 x 1.5

Sky
Charts
8, 9,
10

Apus

(ay-pus) The Bird of Paradise

This faint constellation is directly below Triangulum Australe, the Southern Triangle. Being close to the southern pole, it cannot be seen from most northern latitudes. *Apus* is an ancient Greek word that means "footless" and derived from *Apus Indica*, the name given to India's Bird of Paradise. This magnificent bird was offered as a gift to Europeans, but not before its unsightly legs were cut off.

Apus, as seen in Johann Bode's Uranographia, complete with an early use of constellation boundaries

A 12 inch (300 mm) telescope captured this view of NGC 6101.

🔭 **S Apodis**: This is a "backward" nova. Usually the star shines slightly brighter than magnitude 10, bright enough to see through a small telescope, but at irregular intervals it erupts, possibly sending dark, sootlike material into its atmosphere. It then fades dramatically by some 100 times, to about magnitude 15. After staying faint for several weeks, it slowly returns to its original brightness.

🔭 **Theta (θ) Apodis**: This variable star ranges from magnitude 6.4 to below 8 in a semi-regular cycle over 100 days.

🔭 **NGC 6101**: A faint globular cluster, large and slightly irregular, NGC 6101 can be seen as a small, misty spot through a small telescope.

135

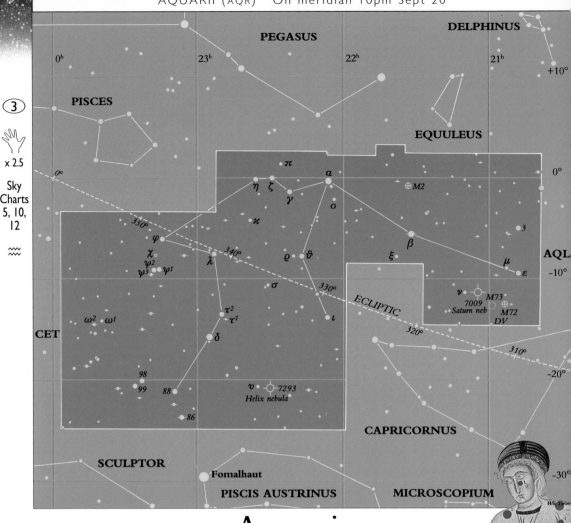

PEGASUS

DELPHINUS

PISCES

EQUULEUS

0°

M2

π

α

η ζ

γ

ο

κ

3

φ

β

AQL

χ
ψ2
ψ3 ψ1

λ

340°

ρ ϑ

ξ

μ

ε

-10°

σ

330°

ι

ν

7009
Saturn neb

M73

M72
DV

ECLIPTIC

320°

310°

CET

ω2 ω1

τ2
τ1

δ

-20°

98
99   88

υ   7293
Helix nebula

310°

86

CAPRICORNUS

-30°

SCULPTOR

Fomalhaut

PISCIS AUSTRINUS

MICROSCOPIUM

Wil Tirion

3

× 2.5

Sky
Charts
5, 10,
12

Aquarius

(ah-KWAIR-ee-us) The Water Bearer

The Water Bearer dates as far back as Babylonian times and is appropriately placed in the sky not far from a dolphin, a river, a sea serpent, and a fish. Of its many mythological associations, it was at times identified with Zeus pouring the waters of life down from the heavens.

Aquarius, the Water Bearer, from a thirteenth-century Italian manuscript

M 2: This fine globular cluster appears as a fuzzy spot of light through binoculars and small telescopes. It is possible, however, to see the cluster's mottled appearance through a 4 inch (100 mm) telescope, and to resolve it into stars through a 6 inch (150 mm) telescope.

The Helix Nebula, at 450 light years away, is the nearest planetary nebula to Earth.

The Saturn Nebula (NGC 7009): This small planetary nebula was named by Lord Rosse who, with his large reflecting telescope, first saw the protruding rays that made it look like a dim version of Saturn with its rings. It is visible through a telescope as a greenish point of light.

The Helix Nebula (NGC 7293): The largest and closest of the planetary nebulae, this cloud takes up half the angular diameter of the Moon in the sky. Because its brightness is spread over a large area, it appears best with a low-power, wide-field telescope or binoculars under a dark sky.

Delta (δ) Aquarids: This strong meteor shower peaks every year on July 28.

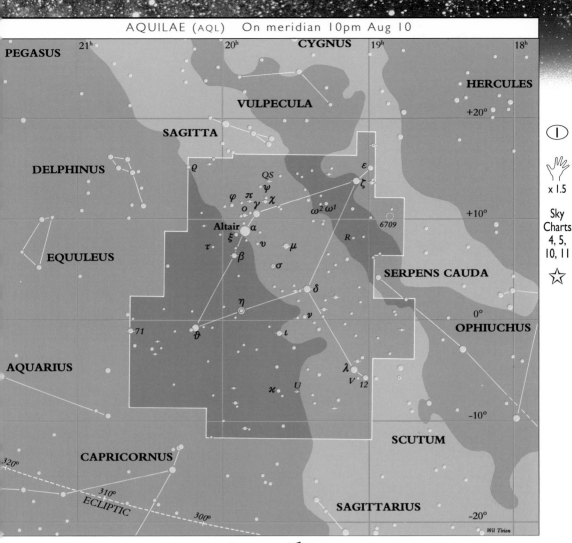

PEGASUS 21ʰ CYGNUS 20ʰ 19ʰ 18ʰ

HERCULES +20°

VULPECULA

SAGITTA

DELPHINUS

ϱ QS ε ζ

φ π ψ χ

o γ ω² ω¹ 6709

Altair α +10°

ξ ν R

EQUULEUS τ μ

β

σ SERPENS CAUDA

δ

η 0°

ν OPHIUCHUS

71 ϑ ι

AQUARIUS λ V 12

ϰ U -10°

SCUTUM

CAPRICORNUS

320°

310°

ECLIPTIC 300° SAGITTARIUS -20°

© Wil Tirion

①

🖐 x 1.5

Sky
Charts
4, 5,
10, 11

☆

Aquila

(uh-KWI-luh) The Eagle

I dentified as an eagle by astronomers of the Euphrates Basin, the constellation of Aquila takes its name from the bird that belonged to the Greek god Zeus. Aquila's main accomplishment was to bring the handsome mortal youth Ganymede to the sky to serve as his master's cup bearer.

Two major novae have appeared in Aquila. The first, in AD 389, was as bright as Venus, and the second, as recently as 1918, was brighter than **Altair**, Aquila's brightest star. One of the brightest stars in the whole sky, Altair is a prominent beacon on the Milky Way between Sagittarius and Cygnus.

Eta (η) Aquilae: This supergiant star is a bright Cepheid variable that changes over a magnitude in brightness (3.5 to 4.4) in a period of little more than a week. At its brightest it rivals **Delta (δ) Aquilae**, and it fades to about the magnitude of **Iota (ι) Aquilae**.

R Aquilae: This Mira star varies in magnitude from 6 to 11.5 over a period of 284 days.

Aquila, with brilliant Altair, astride the Milky Way

NGC 6709: This pretty open cluster consists of a group of closely knit stars against an already rich background of stars. On November 13, 1984, I discovered Comet Levy-Rudenko as a faint fuzzy object in the same field of view as this cluster. The pair made an awesome sight.

137

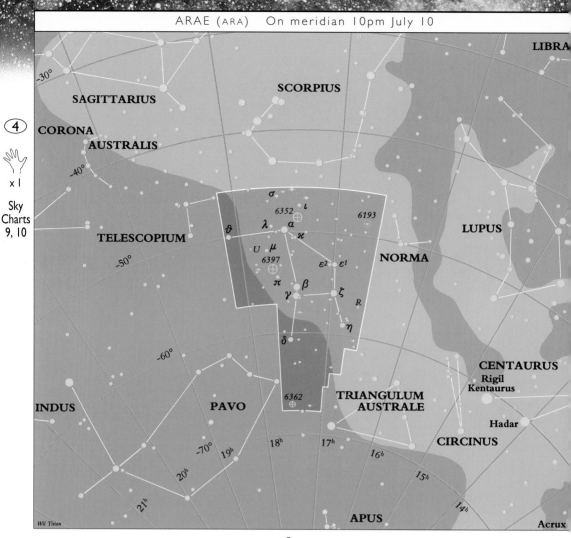

Sky
Charts
9, 10

Ara

(AR-uh) The Altar

Located south of Scorpius, Ara's original Latin name was Ara Centauri—the altar of the centaur Chiron. Half man and half horse, Chiron was thought to be the wisest creature on Earth. Ara has also been referred to as the altar of Dionysus; the altar built by Noah after the flood; the altar built by Moses; and even the one from Solomon's Temple.

U Arae: This Mira-type variable is bright enough to be seen through a small telescope when it is at its maximum of magnitude 8. However, it then drops a full five magnitudes before rising again over a period of more than seven months.

NGC 6397: Possibly the closest globular star cluster to us, this bright cluster is placed between **Beta (β) Arae** and **Theta (θ) Arae**.

The cluster is relatively loose, so an observer with powerful binoculars should be able to detect it without difficulty and perhaps even resolve its faint stars. It may be as large as 50 light years across.

Ara, alongside the Milky Way, with NGC 6397 clearly seen between Beta (β) and Theta (θ) Arae

(Above) Ara, the Altar, as depicted in the 1723 edition of Johann Bayer's Uranometria

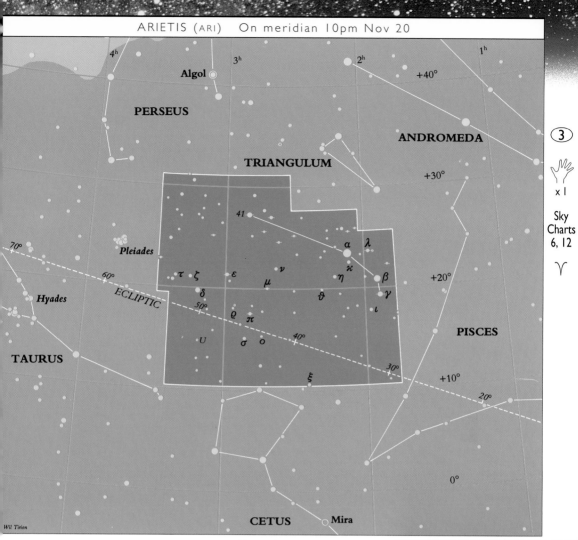

PERSEUS

ANDROMEDA

+40°

Algol

TRIANGULUM

+30°

41

α λ

Pleiades

τ ζ   ε   ν   ϰ   β   +20°

μ   η   γ

ECLIPTIC   δ   ϑ   ι

Hyades   50°   ρ   π

U   σ ο   40°

TAURUS   PISCES

ξ   30°   +10°

20°

CETUS   Mira   0°

Wil Tirion

Aries

(AIR-eez) The Ram

The ancient Babylonians, Egyptians, Persians, and Greeks all called this group of stars the Ram. In one version of the Greek legend, the king of Thessaly had two children, Phrixus and Helle, who were abused by their stepmother. The god Hermes sent a ram with a golden fleece to carry them to safety on its back. Helle fell off the ram as it was flying across the strait that divides Europe from Asia, a body of water the Greeks called the Hellespont, the sea of Helle (now the Dardanelles). Phrixus was carried to safety on the shores of the Black Sea, where he sacrificed the ram and its fleece was placed in the care of a sleepless dragon. It was from here that Jason and the Argonauts stole it.

The brighter stars of Aries, with the stars of Triangulum lying to the north

Aries is the zodiac's first constellation, since the Sun at one time was entering Aries on the day of the vernal equinox—the moment when it crosses from the southern to the northern half of the celestial sphere. However, because of the Earth's precession, the Sun is now in Pisces at the vernal equinox.

Aries is well known and is not difficult to find, but it has few objects of interest.

Gamma (γ) Arietis: In 1664, Robert Hooke was following the motion of a comet when he chanced upon this beautiful double star. One of the earliest doubles to be found with a telescope, this star has a separation of 8 arc seconds, and is easy to find and observe.

A thirteenth-century Italian view of Aries

139

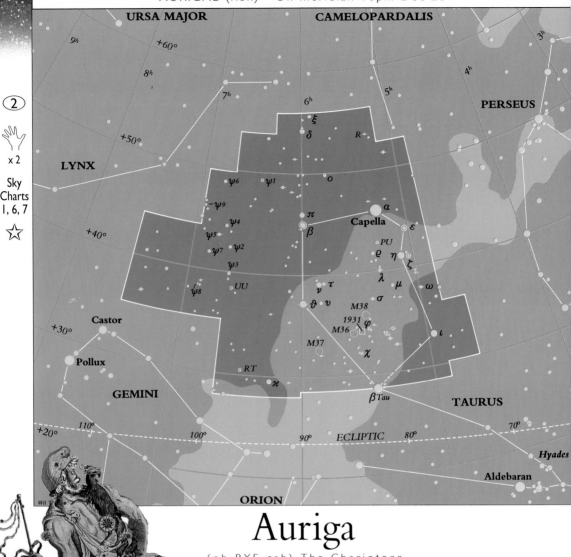

URSA MAJOR   CAMELOPARDALIS

9ʰ   +60°

8ʰ   4ʰ

7ʰ   6ʰ   5ʰ   PERSEUS

②   +50°   ξ
                         δ   R
𝖂 x 2
LYNX   ο
Sky
Charts   ψ6   ψ1
1, 6, 7   ψ9   π   α
              +40°   ψ4   β   Capella   ε
☆   ψ5   PU
              ψ7 ψ2   ϱ η ζ
                 ψ3   λ μ
                              ν τ   ω
              ψ8   UU   ϑ υ   σ
              Castor   M38
              +30°   1931   φ
                             M36   ι
Pollux   M37   χ
              GEMINI   RT
                             ϰ   βTau   TAURUS
              +20°   70°
              110°   100°   90° ECLIPTIC 80°
                                                Hyades
                                          Aldebaran
Wil T   ORION

Auriga

(oh-RYE-gah) The Charioteer

This lovely multi-sided figure is easy to find in the sky, largely because of bright Capella, the she-goat star, and her retinue of three little kids. Ancient legends portray Auriga as a charioteer carrying a goat on his shoulder and two or three kids on his arm. The charioteer is also seen as Erechtheus, the son of Hephaestus (the Roman god Vulcan), who invented a chariot to move his crippled body about.

Capella has been seen as the she-goat star since Roman times. Almost 50 light years away, Capella is similar to our Sun, only larger.

👁 **Epsilon (ε) Aurigae**: An extraordinary variable system, this supergiant star fades when its companion passes in front of it once every 27 years. During an eclipse, its brightness drops by two-thirds of a magnitude. The deepest phase of the eclipse lasts a full year, which may indicate that the companion is surrounded by an enormous disk of gas and dust.

M 36: This bright open star cluster is some 5 degrees southwest of **Theta (θ) Aurigae**, and contains about 60 stars of 8th magnitude and fainter.

M 37: This is an exceptional open star cluster, almost the size of the Moon, and one of the finest in the northern sky. Binoculars will show this cluster as a misty spot. A small telescope will reveal its large number of stars.

M38: This small cluster of stars resembles the Greek letter π (pi) when seen in a small telescope.

(Above) Auriga, with brilliant Capella on his back, as seen in a seventeenth-century atlas

Lying 4,600 light years away, M 37 still presents a magnificent spectacle.

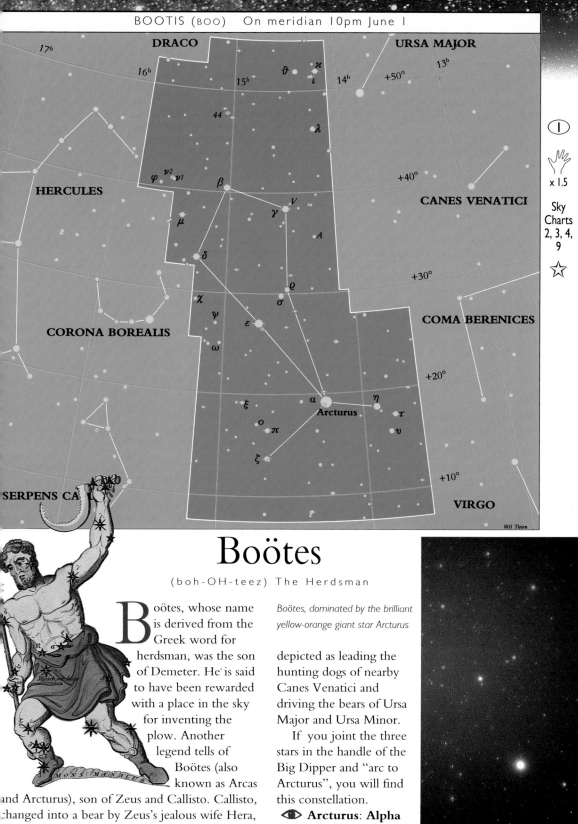

DRACO

URSA MAJOR

17ʰ

16ʰ

15ʰ

14ʰ

13ʰ

+50°

44

λ

HERCULES

φ ν2 ν1

β

+40°

CANES VENATICI

μ

V

γ

A

δ

+30°

χ

ϱ

COMA BERENICES

ψ

σ

ε

CORONA BOREALIS

ω

+20°

ξ

α

η

Arcturus

τ

o

π

υ

ζ

+10°

SERPENS CA

VIRGO

Wil Tirion

Boötes

(boh-OH-teez) The Herdsman

Boötes, whose name is derived from the Greek word for herdsman, was the son of Demeter. He is said to have been rewarded with a place in the sky for inventing the plow. Another legend tells of Boötes (also known as Arcas and Arcturus), son of Zeus and Callisto. Callisto, changed into a bear by Zeus's jealous wife Hera, was almost killed by her son when he was out hunting. Zeus rescued her, taking her into the sky where she became Ursa Major, the Great Bear.

The name Arcturus (the constellation's brightest star) comes from the Greek meaning "guardian of the bear". Sometimes Arcturus is

(Above) The Herdsman, with Arcturus at his knee, as depicted in the Urania's Mirror constellation cards (1825)

Boötes, dominated by the brilliant yellow-orange giant star Arcturus

depicted as leading the hunting dogs of nearby Canes Venatici and driving the bears of Ursa Major and Ursa Minor.

If you joint the three stars in the handle of the Big Dipper and "arc to Arcturus", you will find this constellation.

Arcturus: Alpha (α) Boötis, this yellow–orange star is 37 light years away from us, making it one of the closest of the bright stars. Arcturus's actual position in the sky has changed by over twice the Moon's apparent diameter in the last 2,000 years; astronomers say that Arcturus has a large *proper motion*. I discovered Comet Levy (1987y) near Arcturus one September evening in 1987.

141

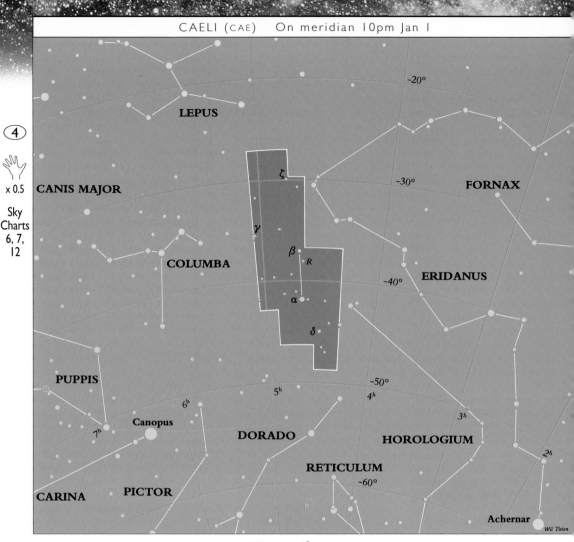

Caelum

(SEE-lum) The Chisel

One of the least conspicuous of all the constellations, Caelum is one of the many regions in the Southern Hemisphere skies that was named by eighteenth-century astronomer Nicolas-Louis de Lacaille. It comprises a largely empty region of the heavens between the constellations of Columba, the Dove, and Eridanus, the River.

R Caeli: A bright Mira-type variable, this star changes from magnitude 6.7 to 13.7 over a period of about 13 months.

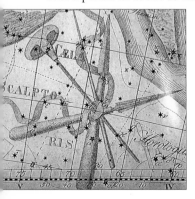

A tiny patch of southern sky, Louis de Lacaille's Caela Sculptoris, the Sculptor's tool, is now known simply as Caelum, a single chisel. Its few faint stars, with a magnitude of 5 at best, could equally well have become part of nearby Columba or Eridanus.

SKYWATCHING TIP

Recording your observations is, in many ways, as important as making them. Whether the notes are in written, sketched, or taped form, the act of recording will enhance your powers of observation. Keep your notes simple and to the point. This is an example of what a Northern Hemisphere observer might record when attempting to locate Caelum:

• Tried to find Caelum, but could see no stars through the horizon haze to the south.

• Made out Alpha (α) and Beta (β) Caeli despite horizon haze.

• Saw bright meteor, magnitude 1. Began near Orion's belt, then went south through Lepus and vanished in Caelum with a bright burst. Greenish in color.

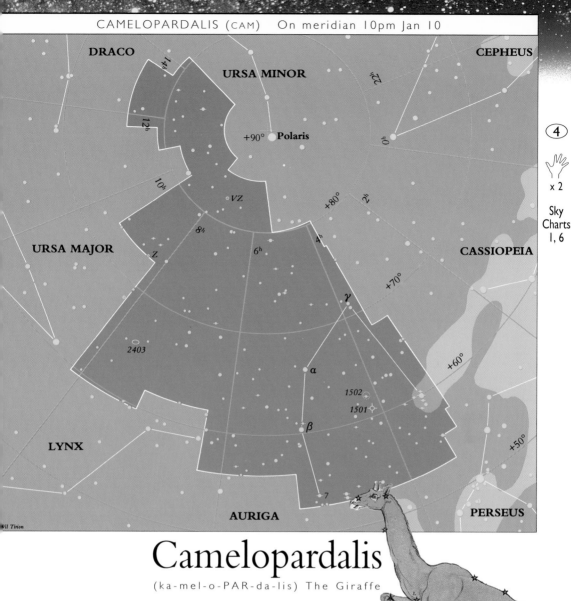

DRACO

URSA MINOR

CEPHEUS

+90° Polaris

VZ

URSA MAJOR

Z

CASSIOPEIA

+80°

+70°

γ

2403

+60°

α

1502

1501

LYNX

β

+50°

7

AURIGA

PERSEUS

Wil Tirion

④

🖐
× 2

Sky
Charts
1, 6

Cameloparalis

(ka-mel-o-PAR-da-lis) The Giraffe

What is a giraffe doing next to two bears and a dragon in the frigid sky near the North Star? Camelopardalis was dreamed up by Bartsch in 1624, who claimed that it represented the camel that brought Rebecca to Isaac. ("Camel-leopard" was the name the Greeks gave to the giraffe, as they thought it had the head of a camel and a

The Giraffe, as featured in the constellation cards of Urania's Mirror (1825)

leopard's spots.) The constellation lies in the large space between Auriga and the bears.

Z Camelopardalis: This cataclysmic variable star erupts every two or three weeks from its minimum of magnitude 13 to a maximum of 9.6, which is still quite faint. Its resemblance to other such variables ceases when, while fading, it stops changing and hovers at an intermediate magnitude. This "standstill" might last for months before the decline resumes. In the late 1970s, Z Cam stayed around magnitude 11.7 for several years.

VZ Camelopardalis: This bright star varies irregularly over the small range between magnitudes 4.8 and 5.2. Located close to Polaris, VZ Cam is visible every night of the year from most northern latitudes.

At 12th magnitude, NGC 1501 is faint for small telescopes, but a 20 inch (500 mm) one shows it with its 14th magnitude star.

143

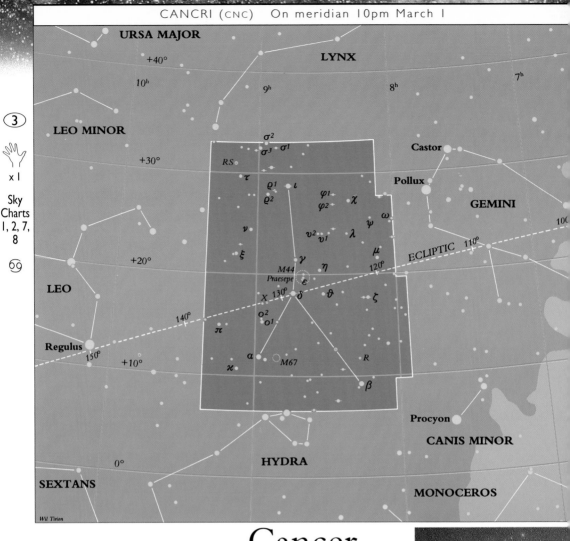

URSA MAJOR

+40°

LYNX

10ʰ   9ʰ   8ʰ   7ʰ

LEO MINOR

+30°   RS

σ²
σ³ σ¹

Castor

Pollux

GEMINI

100

τ

ϱ¹ ι
ϱ²

φ¹
φ²   χ

ψ ω

ν

ν²
ν¹   λ   μ

ECLIPTIC 110°

+20°   ξ

γ   η   120°

M44
Praesepe   ε

ζ

LEO

140°   χ 130° δ ϑ

ο²
ο¹

Regulus   150°   +10°   π

α   M67   R

χ

β

0°

Procyon

CANIS MINOR

SEXTANS

HYDRA

MONOCEROS

Wil Tirion

Sky Charts 1, 2, 7, 8

69

Cancer

(CAN-ser) The Crab

In Greek mythology, Cancer was sent to distract Hercules when he was fighting with the monster Hydra. The crab was crushed by Hercules's foot, but as a reward for its efforts Hera placed it among the stars. The zodiacal symbol represents the crab's claws.

Millennia ago, the Sun reached its summer solstice (its northernmost position in the sky—declination 23.5 degrees north) when it was in front of this constellation. It was then overhead at a northern latitude we call the Tropic of Cancer. As a result of precession, the Sun's most northerly position has now moved westward to the border of Gemini and Taurus.

Cancer lies between Gemini and Leo—two of the sky's showpieces. It has no star brighter than 4th magnitude and its only claims to fame

A sixteenth-century Turkish representation of Cancer

The faint stars of Cancer harbor the open cluster M44— a celestial showpiece.

are its membership of the zodiac and beautiful M 44.

The Praesepe or Beehive (M 44): One of the sky's finest open clusters, this is easy to see through binoculars from the city and with the naked eye from a dark location. There are over 200 stars in the Praesepe. Spread over 1½ degrees, they are best seen with binoculars.

M 67: This open cluster has 500 faint stars spread over ½ degree. Although you can find it with binoculars, your best view will be through a small telescope's low-power eyepiece.

R Cancri: This bright long-period variable is easily visible through binoculars when near its 6.2 magnitude maximum. It varies down to 11.2 and back in almost precisely a year.

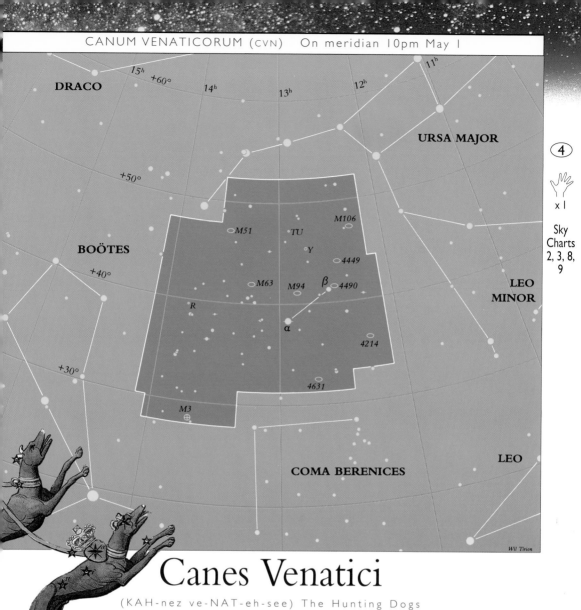

DRACO

15ʰ +60°   14ʰ   13ʰ   12ʰ   11ʰ

URSA MAJOR

④

🖐 x 1

Sky Charts 2, 3, 8, 9

+50°

M106

M51   TU

Y

BOÖTES   4449

+40°   M63   M94   β 4490

R

α

4214

LEO MINOR

+30°   4631

M3 ⊕

LEO

COMA BERENICES

Wil Tirion

Canes Venatici

(KAH-nez ve-NAT-eh-see) The Hunting Dogs

This constellation, tucked away just south of the Big Dipper's handle, contains a wide variety of deep sky objects. Conceived by Hevelius in about 1687, Canes Venatici are the hunting dogs, Asterion and Chara, held on a leash by Boötes as he hunts the skies of the north for the bears Ursa Major and Ursa Minor.

Cor Caroli: The heart of Charles, **Alpha (α) Canum Venaticorum** is believed to have been named by Edmond Halley after his patron, Charles II. It is a wide double (separation 20 arc seconds), easily split by the smallest telescope.

M 3: A rare gem of the northern sky, this globular cluster is midway between Cor Caroli and Arcturus. Some 35,000 light years away and 200 light years across, M 3 begins to resolve into stars through a small telescope.

Y Canum Venaticorum (E-B 364): Named La Superba by Secchi in the nineteenth century, this 5th magnitude star is splendidly red. It varies from magnitude 5.2 to 6.6 over 157 days.

The Whirlpool Galaxy (M 51): This famous galaxy appears as a round, 8th magnitude glow with a bright nucleus. A 12 inch (300 mm) telescope will show its spiral structure.

(Left) The Hunting Dogs were held on a leash by Boötes, the Herdsman, in Urania's Mirror (1825). (Below) The spiral galaxy M 51 and its companion NGC 5195 form one of the best known images in astronomy.

MONOCEROS

ORION

ERI

Rigel

0°

-10°

2360
R

γ

ϑ

μ

Sirius

ι

α

ν3
ν1
ν2

β

π

M41

ξ2
ξ1

-20°

o2
o1

τ 2362

δ

ω

σ

ε Adhara

ζ

η

LEPUS

-30°

PYXIS

x

λ

PUPPIS

8ʰ

7ʰ

6ʰ

COLUMBA

5ʰ

-40°

VELA

CAELUM

Wil Tirion

Canis Major

(KAH-niss MAY-jer) The Great Dog

One of the most striking of all the constellations, the Great Dog is marked by the brilliant star Sirius, commonly known as the Dog Star—the brightest star in the entire sky. Sirius is said to be responsible for the Northern Hemisphere's hot, muggy "dog days" that occur in September. Legend has it that because Sirius rises at the same time as the Sun during late summer, its brightness adds to the Sun's energy, producing additional warmth.

Canis Major and its neighboring constellation, Canis Minor, the Little Dog, appear in quite a number of myths. One legend has the two dogs sitting patiently under a table at which the Twins are dining. The faint stars that can be seen scattered in the sky between Canis Minor and Gemini are the crumbs the Twins have been feeding to the animals.

According to the ancient Greeks, Canis Major could run incredibly fast. Laelaps, as they called him, is said to have won a race against a fox that was the fastest creature in the world. Zeus placed the dog in the sky to celebrate the victory.

Another myth has the Great Dog and Little Dog assisting Orion while he is out hunting, his favorite sport. With his eye fixed on Lepus, the Hare, crouching just below Orion, Canis Major

Canis Major, the Great Dog, as depicted in the Urania's Mirror *constellation cards (1825)*

146

The great star Sirius clearly dominates this view of Canis Major, although the large disk is not seen by eye—it is a photographic artifact due to the star's great brightness.

seems ready to pounce. In other versions of the story, Sirius is Orion's hunting dog.

The ancient Egyptians had a great deal of respect for Sirius. After being close to the Sun for some months, the star would rise just before dawn in late summer, an event known as its heliacal rising. This would herald the annual flooding of the Nile Valley, the waters re-fertilizing the fields with silt. This event was of such importance to them that it marked the beginning of their year.

Sirius: The sky's brightest star, Sirius is only 8.7 light years from Earth. Its great brilliance is also due to its being some 40 times more luminous than the Sun.

In 1834, Friedrich Bessel noted that Sirius had a strange wobble to its position, indicating an unseen companion. In 1862, the famous telescope maker Alvan Clark, while testing a new 18½ inch (460 mm) refractor on Sirius, discovered the faint star we now know as the Pup. It is a white dwarf star, its density being so great that a piece of it the size of this book might weigh around 197 tons (200 tonnes). On its own, the Pup would be a respectable star visible through a telescope at magnitude 8.4, but its closeness to mighty Sirius makes it a difficult target, requiring a telescope of 10 inch (250 mm) aperture and very steady viewing conditions.

In ancient Greek and Roman astronomical records, Sirius is quite frequently described as being "ruddy" or "reddish" in color. Was Sirius red in the recent past? Current thinking sees this as unlikely, since other bright stars were sometimes also described as red. This might perhaps have been because of the colors that can be seen when bright stars twinkle.

M 41: A beautiful open cluster, M 41 is surrounded by a rich field of background stars. If you look at it through a telescope, you will be able to see a distinctly red star near the cluster's center.

NGC 2362: This cluster of several dozen stars is tightly packed around **Tau (τ) Canis Majoris**. What is not clear is whether Tau (τ) is actually a member of the cluster or just a chance foreground star.

M 41 is a large, bright galactic cluster, almost the size of the full Moon.

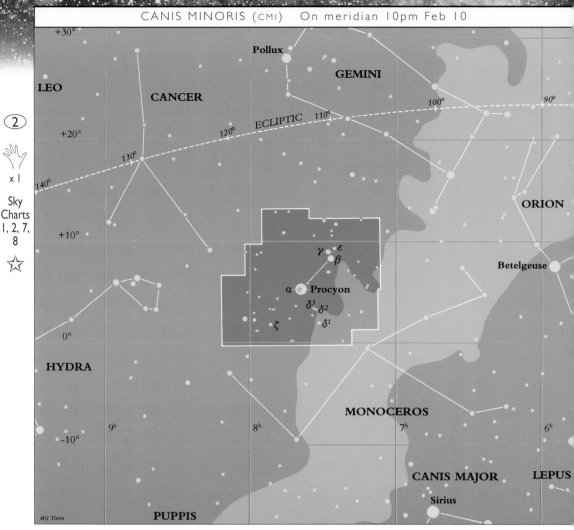

+30°

Pollux

GEMINI

LEO

CANCER

ECLIPTIC

100°

90°

+20°

120°

110°

130°

ORION

+10°

140°

γ   ε
β

Betelgeuse

α ○ Procyon

δ³ δ²

δ¹

ζ

0°

HYDRA

MONOCEROS

-10°

9ʰ

8ʰ

7ʰ

6ʰ

CANIS MAJOR

LEPUS

Sirius

Wil Tirion

PUPPIS

②

ᵛⁿᵐ
x 1

Sky
Charts
1, 2, 7,
8

☆

Canis Minor

(KAH-niss MY-ner) The Little Dog

Canis Minor, Canis Major's playful smaller companion, has only two stars brighter than 5th magnitude—Procyon (Greek for "before the dog", as it rises before Sirius) and Gomeisa. Besides being one of Orion's hunting dogs, Canis Minor was also said to be one of Actaeon's hounds. One day Actaeon surprised Artemis, goddess of the chase and the forests, while she was bathing in a pond with her companions. Spellbound by her great beauty, he paused for a moment and she saw him. Furious that a mortal had seen her naked, Artemis turned him into a stag, set her pack of hounds upon him, and he was devoured.

👁 **Procyon: Alpha (α) Canis Minoris**, this beautiful deep yellow star follows Orion across the sky. Only 11.3 light years away, it is accompanied by a white dwarf that is much fainter than the Pup that accompanies Sirius.

Beta (β) Canis Minoris: This star is set in a beautiful field which includes one quite red star.

Procyon at upper left, Sirius at the bottom, and Orion at right are beacons in the sky early each year.

148

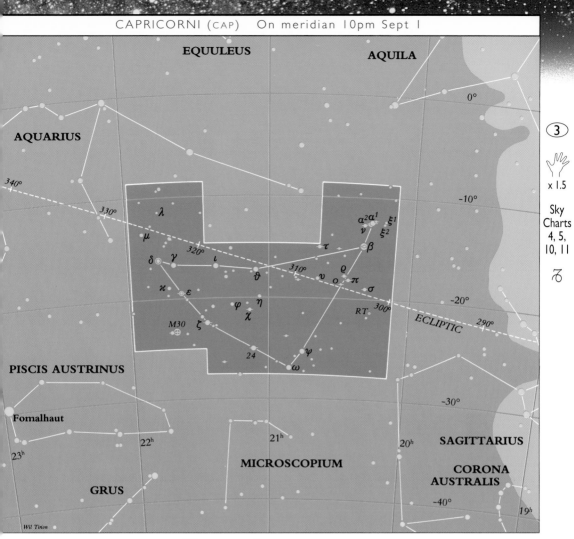

EQUULEUS

AQUILA

AQUARIUS

0°

③

🖐
× 1.5

Sky
Charts
4, 5,
10, 11

♑

-10°

-20°

ECLIPTIC

PISCIS AUSTRINUS

Fomalhaut

-30°

22ʰ

21ʰ

20ʰ

SAGITTARIUS

23ʰ

MICROSCOPIUM

CORONA
AUSTRALIS

GRUS

-40°

19ʰ

Wil Tirion

Capricornus

(kap-reh-KOR-nuss) The Sea Goat, Capricorn

Capricornus has been named for a goat since the time of the Chaldeans and Babylonians. Sometimes it is shown as a goat, but more commonly it is depicted as a goat with the tail of a fish. This might relate to a story about the god Pan, who, when fleeing the monster Typhon, leaped into the Nile. The part of him that was underwater turned into a fish tail, while his top half remained that of a goat.

Several thousand years ago, the Sun reached its southernmost position in the sky (its winter solstice—declination 23.5 degrees south) when it was in front of Capricornus. During this time it was overhead at a southerly latitude we call the Tropic of Capricorn. It still carries this name, although the Sun, as a result of precession, is now in Sagittarius at the time of the winter solstice.

The triangle of stars forming Capricornus is easily recognized, although the stars are no brighter than 3rd magnitude.

Capricornus, the zodiac's least visible constellation, can be found by joining Aquila's three brightest stars in a line southward.

👁 **Alpha (α) Capricorni**: This double has a separation of 6 arc minutes—a naked-eye test for a night's clarity and steadiness. The pair is a double by coincidence, but each star is itself a true binary.

🔭 **M 30**: Perhaps 40,000 light years away, this globular cluster has a fairly dense center. It is not well resolved in small telescopes.

Capricornus, as depicted in a fresco at the Villa Farnese, Caprarola, Italy (1575)

149

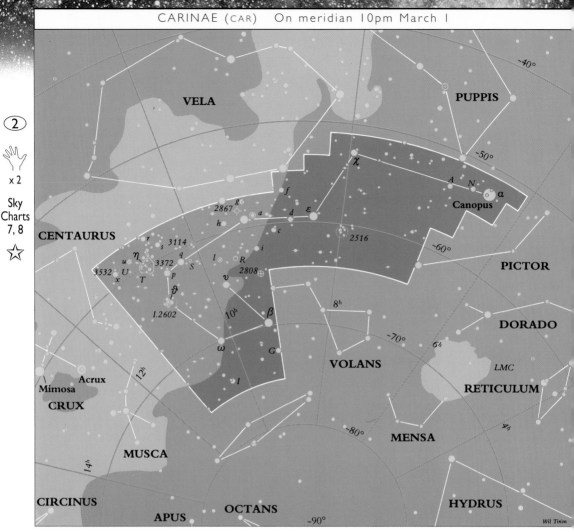

②
🖐 x 2

Sky Charts 7, 8

☆

Carina

(ka-RYE-nah) The Keel

This Southern Hemisphere constellation is in the middle of one of the richest parts of the Milky Way and under a dark sky it is breathtaking. With binoculars, you can see at least half a dozen bright open clusters.

Carina is part of what was once a huge constellation known as Argo Navis, the Ship Argo—the vessel that Jason and his Argonauts sailed in on their search for the Golden Fleece. Argo Navis covered such a rich area of sky that it was divided into four separate constellations: Pyxis, Puppis, Vela, and Carina.

👁 **Canopus: Alpha (α) Carinae**, this yellow

The magnificent Eta (η) Carinae Nebula will reward any size of binoculars or telescope.

supergiant is the second brightest star in the sky and is some 74 light years away.

🔭 **Eta (η) Carinae**: In 1677, Edmond Halley noticed that this star had brightened. In 1827 it shot up to 1st magnitude, and for a few weeks in 1843 it tied with Sirius as the brightest star in the sky. In recent years, however, Eta (η) has been too faint to be seen without binoculars.

The star is famous primarily for the surrounding **Eta (η) Carinae Nebula (NGC 3372)**, the most exquisite nebula in the Milky Way. It is 2 degrees across, with dark rifts appearing to break it up. Superimposed on the brightest part of the nebula is the dark **Keyhole Nebula (NGC 3324)**.

🔭 **NGC 3532**: A brilliant open cluster some 3 degrees from Eta Carinae, this is the finest of the clusters in Carina, with about 150 stars visible in a telescope at low magnification.

🔭 **IC 2602**: This open cluster of scattered, bright stars around **Theta (θ) Carinae** is seen to best advantage in binoculars or in the eyepiece of a wide-field telescope.

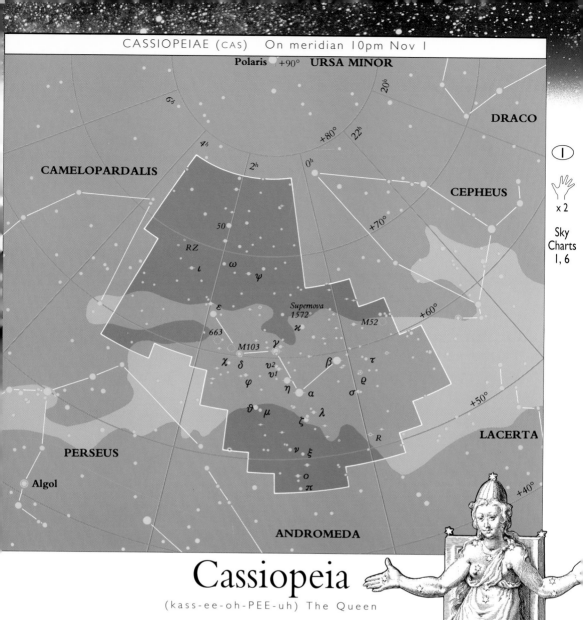

Polaris +90° **URSA MINOR**

DRACO

CAMELOPARDALIS

CEPHEUS

50
RZ
ι   ω
ψ
ε
663
M103   γ
χ δ   v²
φ   v¹
η
ϑ   μ   ζ   λ
ν ξ

Supernova
1572
ϰ

M52

β   τ
α   ϱ
σ

R

LACERTA

PERSEUS

Algol

ANDROMEDA

+80°
+70°
+60°
+50°
+40°

I
x 2

Sky
Charts
1, 6

Cassiopeia

(kass-ee-oh-PEE-uh) The Queen

This striking W-shaped figure is on the other side of Polaris from the Big Dipper. Most prominent in the Northern Hemisphere's winter sky, Cassiopeia is visible all year from mid-northern latitudes. In Greek mythology, she was queen of the ancient kingdom of Æthiopia—wife of Cepheus and mother of Andromeda.

The Romans saw Cassiopeia as having been chained to her throne, as a punishment for her boastfulness, and placed in the heavens to sometimes hang upside down. Arab cultures pictured the constellation as a kneeling camel.

An engraving of Cassiopeia by Jacob de Gheyn (1621)

👁 **Gamma (γ) Cassiopeiae:** This star lies at

The prominent "W" of Cassiopeia is an unmistakable feature in the northern sky.

the center of Cassiopeia's W figure. Normally the constellation's third brightest star, it is an irregular variable. Over a few weeks in 1937, it was the brightest star in the constellation and almost as bright as Deneb in Cygnus. Known as a shell star, Gamma (γ) Cassiopeiae is slowly losing mass into a disk or shell that surrounds it, and alterations in the shell's thickness might be responsible for its variations in brightness.

🔭 **M 52:** This group of about 100 stars is one of the richest in the northern half of the sky, but only one of several open clusters scattered throughout Cassiopeia.

🔭 **NGC 663:** A small open cluster of quite faint stars, NGC 663 is an attractive sight in a small telescope.

151

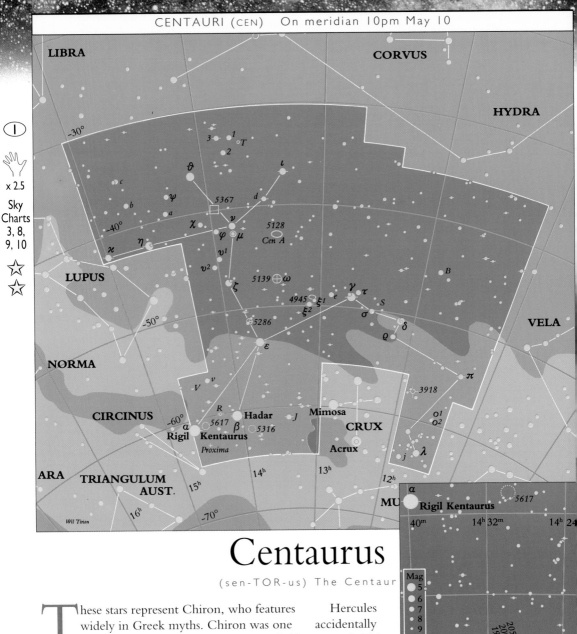

LIBRA

CORVUS

HYDRA

x 2.5
Sky
Charts
3, 8,
9, 10

LUPUS

VELA

NORMA

CIRCINUS

Hadar

Mimosa

CRUX

Rigil
Kentaurus

Proxima

Acrux

ARA

TRIANGULUM
AUST.

MU

Wil Tirion

Rigil Kentaurus

Proxima
Centauri

Mag

Centaurus

(sen-TOR-us) The Centaur

These stars represent Chiron, who features widely in Greek myths. Chiron was one of the Centaurs—creatures that were half man, half horse. Unlike the other Centaurs, who were monstrous and brutal, Chiron was extremely wise, and tutored such humans as Jason and Hercules.

Hercules accidentally wounded him, and Chiron, in great pain but unable to die because he was immortal, pleaded with the gods to end his suffering. Zeus mercifully allowed Chiron to die, and placed him among the stars.

This huge constellation includes the bright Milky Way near the Southern Cross (Crux).

Alpha (α) Centauri: At the foot of the Centaur, this star is only 4.3 light years away and is the Sun's nearest neighbor. One of the prettiest binary stars, its two components revolve around each other once every 80 years. The separation is currently around 20 arc seconds, but this will close to 2 arc seconds by about the year 2035. Alpha (α) and **Beta (β) Centauri** are the bright "pointers" to the Southern Cross.

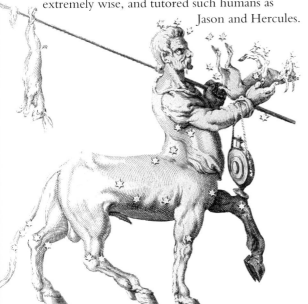

The Centaur, from an engraving by Jacob de Gheyn, (1621)

(Above) A chart of the region near Proxima Centauri

The Southern Cross, in the lower right corner, is almost surrounded and dwarfed by Centaurus.

Proxima Centauri: In 1915, R.T. Innes was measuring the proper motions of stars around Alpha (α) Centauri when he found a faint, 10.7 magnitude star moving at the same rate and in the same direction as the two stars of Alpha (α) Centauri but about 2 degrees away. A very small red dwarf star only 40,000 miles (25,000 km) across, this star is actually a little closer to us than the other two, but it is presumed to be their companion. It flares occasionally, jumping by half a magnitude or more, usually returning to its normal brightness within half an hour.

Omega (ω) Centauri: This globular cluster is, by most accounts, the finest example in the entire sky, with perhaps 1 million members. With the naked eye it is visible as a fuzzy star of 4th magnitude, so bright that Johann Bayer, in the early seventeenth century, gave it the designation Omega (ω). Only 17,000 light years away, this is one of the closest clusters to us, second only to **NGC 6397** in Ara. Unlike most globulars, it is oval rather than round in shape.

Omega (ω) Centauri is a spectacular sight in any telescope over about 6 inches (150 mm).

A 3 inch (75 mm) telescope will show a large, fuzzy disk with mottled edges, while a 6 inch (150 mm) one will resolve it into stars. Viewed in an even larger telescope, under a dark sky, it looks magnificent, with the field of a low-power eyepiece overflowing with faint stars.

NGC 5128: Located only 4½ degrees north of Omega (ω) Centauri, this peculiar elliptical galaxy is distinguished by a strange dark band that crosses its center—probably the result of a collision with a spiral galaxy. A strong source of radio energy, known to radio astronomers as **Centaurus A**, this galaxy emits more than 1,000 times the radio energy of our own galaxy. The dark dust lane is apparent in dark skies with a 4 inch (100 mm) or larger telescope.

NGC 3918: This is a planetary nebula not far from the Southern Cross, presenting a classic blue-green disk about 12 arc minutes across, like a larger version of Uranus.

The green star seen here in the dust lane of NGC 5128 is a supernova that was bright only when the green image of this red/green/blue composite was photographed.

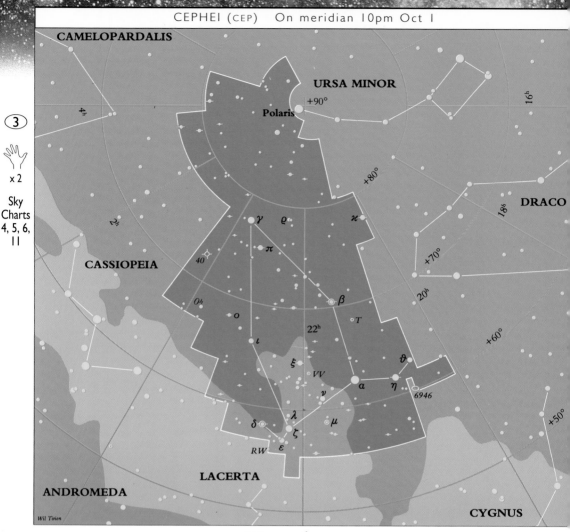

CAMELOPARDALIS

URSA MINOR
+90°
Polaris

DRACO

CASSIOPEIA

γ ϱ $\varkappa$

π

40

0^h

o

β

T

22^h

ι

ξ

VV

ϑ

ν α η

6946

λ

δ

μ

ζ

RW ε

+80°

+70°

20^h

+60°

+50°

18^h

16^h

2^h

4^h

3

x 2

Sky
Charts
4, 5, 6,
11

ANDROMEDA

LACERTA

CYGNUS

Wil Tirion

Cepheus

(SEE-fee-us) The King

(Above) NGC 6446 is a 9th magnitude spiral galaxy, but a
16 inch (400 mm) telescope will reveal the spiral arms.

King of the ancient land of Æthiopia,
Cepheus was the husband of Cassiopeia
and father of Andromeda. His wife and
daughter are both represented by constellations,
and an account of the myth in which he is
involved is given under Andromeda (p. 132).

Cepheus is an inconspicuous constellation.
The constellation's five bright stars are easy to
find, only because they face the open side of the
W shape of Cassiopeia. It looks a little like a
house with a pointed roof. Although the top of
the roof does not really point to Polaris, it offers
the general direction to the pole at a time of year
when the pointer stars of the Big Dipper are not
readily accessible.

👁 **Delta (δ) Cephei**: One of the most
famous of the variable stars, and the prototype
for the Cepheid variables, Delta (δ) Cephei's
variation was discovered by John Goodricke,
a deaf-mute teenager, in 1784. Its highest
magnitude is 3.5, as bright as neighboring **Zeta
(ζ) Cephei**, and it fades to 4.4, the brightness
of **Epsilon (ε) Cephei**. It completes a cycle
every 5.4 days.

👁 **Mu (μ) Cephei**: This star is so strikingly
red that William Herschel called it the Garnet
Star. Using Zeta (ζ) and Epsilon (ε) as
comparison stars, you can watch it vary in
brightness irregularly over hundreds of days.

154

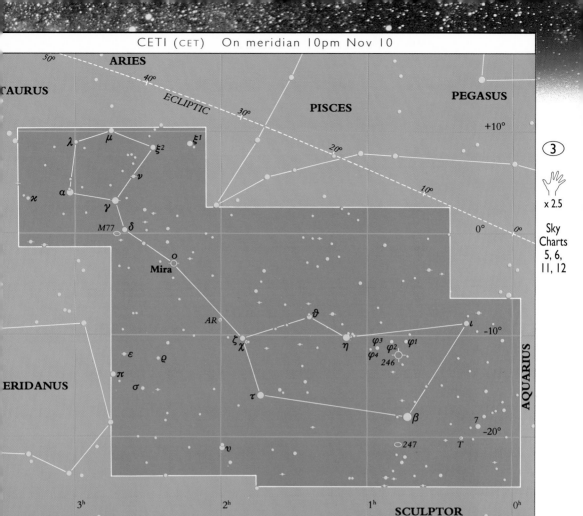

ARIES

TAURUS

ECLIPTIC

PISCES

PEGASUS

50°

40°

30°

20°

10°

+10°

0°

-10°

-20°

-30°

λ

μ

ξ²

ξ¹

ν

κ

α

γ

M77 δ

ο

Mira

AR

ζ χ

ϑ

η

φ³

φ⁴

φ²

246

φ¹

ι

ε

ϱ

π

σ

τ

β

7

247

T

υ

ERIDANUS

AQUARIUS

3ʰ

2ʰ

1ʰ

0ʰ

SCULPTOR

FORNAX

Al Tirion

③

x 2.5

Sky
Charts
5, 6,
11, 12

Cetus

(SEE-tus) The Whale, The Sea Monster

Known by the ancient Greeks as the monster that was about to attack Andromeda when Perseus destroyed it, Cetus was later thought to represent the whale that consumed Jonah. Cetus consists of faint stars, but it occupies a large area of sky. His head is a group of stars not far from Taurus and Aries, and his body and tail lie towards Aquarius.

👁 **Mira: Omicron (ο) Ceti**, known as Mira, is the most famous long-period variable of all. In August 13, 1596, David Fabricius, a Dutch sky-watcher, noticed a new star in Cetus. Over the following weeks, it faded, disappeared, then reappeared in 1609. In 1662 Johannes Hevelius named it Mira Stella, the Wonderful Star. Mira

varies from a magnitude of about 3.4 to a minimum of 9.3 over 11 months.

🔭 **M 77**: The brightest of several galaxies in Cetus, M 77 is a 9th magnitude spiral galaxy with a bright core. A 4 inch (100 mm) telescope shows a faint circular disk around the core.

Mira appears near maximum brightness in this view of Cetus.

(above) Bayer's depiction of Cetus, from his Uranometria (1603), in which Greek letters were introduced to label the stars

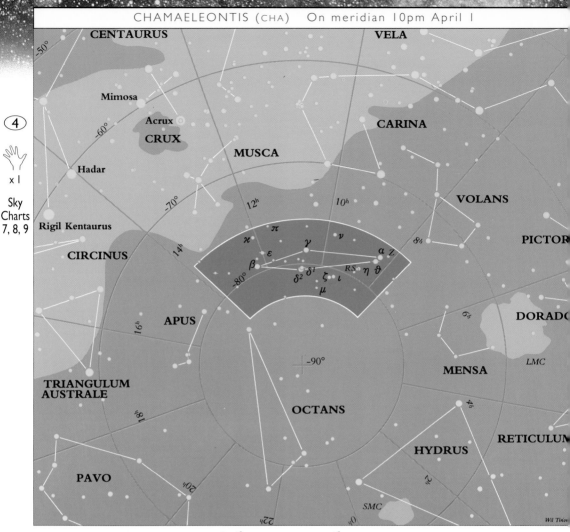

CENTAURUS   VELA

Mimosa

CARINA

Acrux
CRUX

MUSCA

Hadar   VOLANS

Rigil Kentaurus   PICTOR

CIRCINUS

DORADO

APUS   LMC

MENSA

TRIANGULUM
AUSTRALE

OCTANS

RETICULUM

HYDRUS

PAVO

SMC

Wil Tirion

Chamaeleon

(ka-MEE-lee-un) The Chameleon

Johann Bayer drew this constellation early in the seventeenth century, following descriptions that had been given by certain early south sea explorers.

The chameleon is a small lizard found in Africa that can change color to match its surroundings.

Modern researchers have found that the animal changes its color in response to sudden changes light and temperature, and to emotional shock.

One of the smallest and least conspicuous of the constellations, Chamaeleon does a good job hiding in the sky too. Consisting of a few faint stars, it lies close to the south celestial pole, sou of Carina and right beside the south polar constellation of Octans.

Z Chamaeleontis: This faint variable sta erupts periodically. At its minimum it shir at magnitude 16.2, invisible except in 12 inch (300 mm) or larger telescopes. However, every three to four months it undergoes an outburst, rising within a few hours to about magnitude 11.5, and for a few days it is visible through a 6 inch (150 mm) telescope. Even so, this does constitute an easy target in a corner of the sky with few stars.

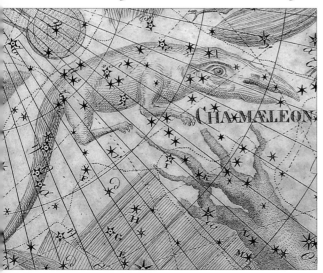

The Chameleon, as depicted by Johann Bode in his Uranographia (1801), extends beyond the modern boundarie of the constellation.

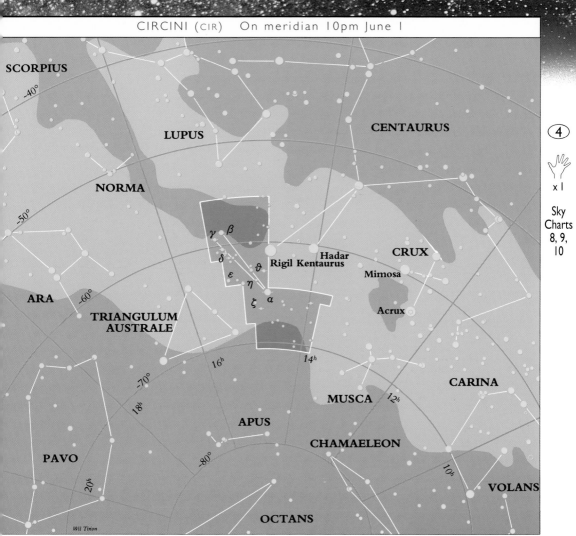

Circinus

(SUR-seh-nus) The Drawing Compass

The early explorers in the south seas were less interested in mythology than in the modern instruments that they relied on to find their way around uncharted waters. The Drawing Compass is one of a number of obscure constellations that were designated by the

Johann Bode's Uranographia *(1801) was the first reasonably complete atlas of stars that can be seen with the naked eye, as can be seen by comparing his map of Circinus with the modern map.*

French astronomer Nicolas-Louis de Lacaille. He worked at an observatory at the Cape of Good Hope from 1750 to 1754, where he compiled a catalogue of more than 10,000 stars.

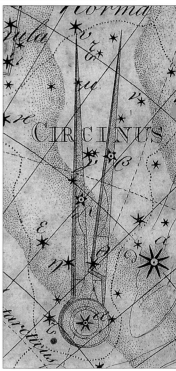

SKYWATCHING TIP

Scale is one of the hardest things to become accustomed to when observing the night sky. If, in your early attempts, you are unable to find a single constellation it might be because you have no idea what size pattern you are looking for.

Firstly, look at the edge of the constellation chart to see how many hand spans will cover the constellation from east to west. Then compare the new constellation chart with some familiar ones.

When using the starfinder charts at the start of this chapter, bear in mind that star groups at the edges of charts will appear larger than they really are, because of distortion.

Alpha (α) Circini: This, at only 3rd magnitude, is the constellation's brightest star. It lies just near the much brighter Alpha (α) Centauri. It is about 65 light years away and has a faint 9th magnitude companion.

157

ORION

Sirius

LEPUS

CANIS MAJOR

ERIDANUS

~10°

~20°

ν1
ν2

σ
μ

δ
λ   α
T

κ   γ
ε   ο

SX   ϑ   ξ   β

CAELUM

~30°

1851

~40°

PUPPIS

π2 π1   η

HOROLOGIUM

~50°

7ʰ   Canopus   6ʰ   5ʰ   4ʰ

8ʰ

PICTOR

CARINA

DORADO

VELA

Wil Tirion

3

x 1

Sky charts 1, 7, 12

Columba

(koh-LUM-bah) The Dove

Immediately south of Canis Major, Columba is a modern constellation named by Petrus Plancius, a sixteenth-century Dutch theologian and mapmaker. This inconspicuous group of stars honors the dove that Noah sent out from the ark after the rains had stopped, to see if it could find dry land.

T Columbae: A Mira variable, this star has a maximum magnitude of 6.7. It drops to magnitude 12.6 and then rises again over a period of seven and a half months.

NGC 1851: Bright and large, this 7th magnitude globular cluster appears as a misty spot through binoculars under a good sky. A 6 inch (150 mm) telescope will begin to resolve the cluster's brightest stars.

Johann Bode's drawing of the Dove in Uranographia (1801) coincides quite closely with the official extent of Columba. (Left) About 11 arc minutes across, NGC 1851 is quite a large globular cluster.

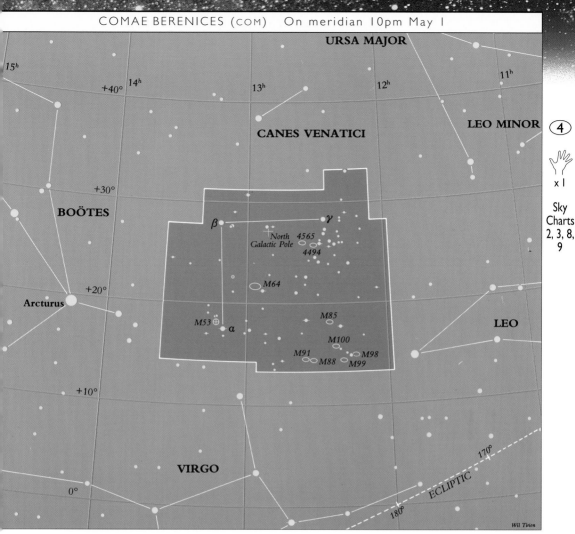

URSA MAJOR

CANES VENATICI

LEO MINOR

④

🖐
x 1

Sky
Charts
2, 3, 8,
9

BOÖTES

North   4565
Galactic Pole
4494

β

γ

M64

Arcturus

M53 ⊕

α

M85

LEO

M100

M91   M98
M88   M99

VIRGO

ECLIPTIC

Wil Tirion

Coma Berenices

(KOH-mah bear-eh-NEE-seez) Berenice's Hair

Between Arcturus and Denebola (Beta [β] Leonis), Coma Berenices has no bright stars and is hard to distinguish, but it is a remarkable area of sky. It is a sprinkling of faint stars superimposed on a cloud of galaxies—the northern end of the Virgo cluster of galaxies. Fainter still, beyond the range of most amateur astronomers, is the Coma cluster of galaxies.

The story behind the constellation is, this time, about real people. Berenice, the beautiful wife of the ancient Egyptian king Ptolemy III, promised to sacrifice her long golden hair to Aphrodite if her husband returned safely from battle, which he did. Her hair was placed in the temple, but it disappeared. The king was about to put the temple guards to death when the court astronomer announced that Aphrodite, delighted with the gift, had placed it in the sky for all to admire.

Berenice's hair, committed to the stars, from Urania's Mirror (1825)

The Blackeye spiral galaxy with its dark dust lane

M 53: This fine globular cluster is about 3 arc minutes in diameter and is located close to **Alpha (α) Comae Berenices.**

The Blackeye Galaxy (M 64): This is one of the most unusual galaxies in the sky. It looks like an ordinary spiral galaxy, with tightly wound arms, but viewed with a 4 to 6 inch (100 to 150 mm) telescope or larger, a huge cloud of dust can be seen dominating its center, giving it the look of a black eye.

NGC 4565: Under a dark sky, a small telescope should show this faint object as a pencil-thin line of haze. It is a spiral galaxy seen edge-on, with a dust lane that becomes apparent in 8 inch (200 mm) telescopes.

159

SERPENS CAUDA

CAPRICORNUS

300°　　290°　　280°　　270°　　~20°
ECLIPTIC　　　　　　　　　　　　　260°

OPHIUCHUS
~30°

SAGITTARIUS

SCORPIUS

6726/7/9

Shaula

γ
α　ε　λ　κ
β
δ　　μ
ζ　η²　ϑ　　-40°
η¹
6541

-50° 17ʰ

19ʰ　　18ʰ

20ʰ

21ʰ　　　　　　　　16ʰ

NORMA

ARA

-60°

INDUS

Wil Tirion

Corona Australis

(kor-OH-nah os-TRAH-lis) The Southern Crown

CORONAE AUSTRALIS (CRA)

O ne of the 48 original constellations
catalogued by Ptolemy in the second
century AD, this small semicircular
group of faint stars is inconspicuous, especially
from the Northern Hemisphere. It lies just south
of Sagittarius and is said to represent a crown

*In Uranometria (1603), Bayer's Southern Crown is a crown of
laurels, as given to victorious athletes in ancient times.*

of laurel or olive leaves. One story has it that the
crown belongs to Chiron.

Another story relating to the crown comes
from Ovid's *Metamorphoses*. Juno discovered that
her husband, Jupiter, was the lover of Semele, a
human. Masquerading as Semele's maid, Juno
suggested that Semele ask Jupiter to appear
before her in all his glory. Jupiter was appalled at
her request, but did not refuse it. When she saw
him in his splendor she was consumed by fire.
Her unborn child was saved, however, to
become Bacchus, god of wine, who honored his
mother by placing the crown in the sky.

NGC 6541: This globular cluster presents
a small nebulous disk to smaller telescopes.
An 8 inch (200 mm) telescope only begins to
resolve the edge into stars.

*NGC 6726–7 at the top, and comet-like NGC 6729 just below
it, lie almost on the border with Sagittarius*

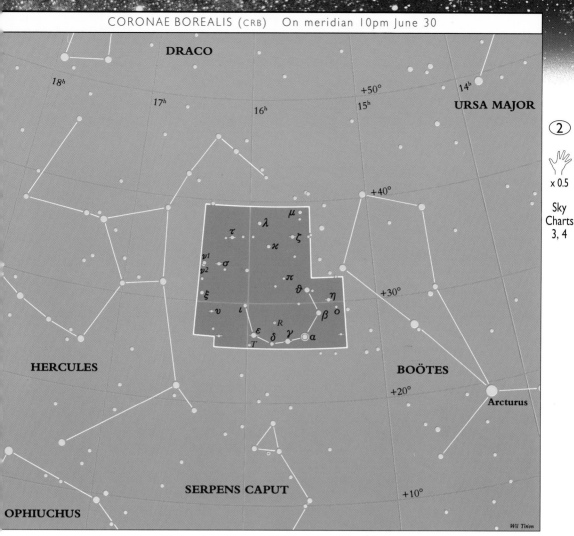

DRACO

18ʰ

17ʰ

16ʰ

15ʰ

14ʰ

+50°

URSA MAJOR

+40°

μ

λ

τ

ζ

ϰ

ν1

σ

ν2

π

ϑ

ξ

η

β

ο

+30°

υ

ι

R

ε

δ

γ

α

T

+20°

HERCULES

BOÖTES

Arcturus

②

✋ x 0.5

Sky
Charts
3, 4

SERPENS CAPUT

+10°

OPHIUCHUS

Wil Tirion

Corona Borealis

(kor-OH-nah bor-ee-AL-is) The Northern Crown

Just 20 degrees northeast of Arcturus lies the Northern Crown, a small semicircle of stars that are faint but very distinct.

There are legends from many cultures to explain its presence in the sky. The Greek myth claims the crown belongs to Ariadne, daughter of Minos, King of Crete. Ariadne was reluctant to accept a marriage proposal from Dionysus (who was in mortal form), since she did not wish to marry a mortal after being deserted by Theseus. To prove he was a god, Dionysus took off his crown and threw it into the heavens as a tribute to her. Satisfied, Ariadne married him and became immortal herself.

A small but clear semicircle of faint stars makes Corona Borealis quite easy to recognize.

R Coronae Borealis: One of the more remarkable stars in the sky, R Cor Bor, as it is generally known, is a nova in reverse. Normally shining at magnitude 5.9, at completely irregular intervals the star will

The Northern Crown appears as a royal crown in Urania's Mirror *(1825).*

suddenly fade, sometimes by as much as 8 magnitudes, as dark material erupts in its atmosphere. It then slowly recovers as the material dissipates.

T Coronae Borealis: Now shining at magnitude 10.2, in 1866 this star suddenly rose to magnitude 2. Known as a recurrent nova, the star repeated the performance unexpectedly in 1946, and will probably do so again.

161

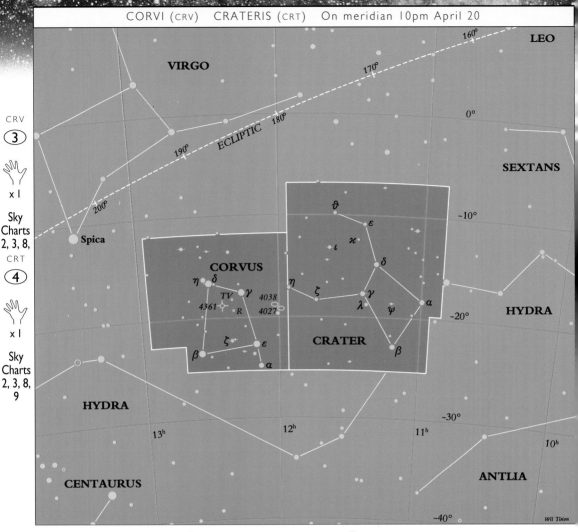

LEO

VIRGO

160°

170°

ECLIPTIC   180°

0°

190°

SEXTANS

200°

-10°

Spica

ϑ

ε

ι ϰ

CORVUS

η δ

δ

η ζ

SEXTANS

TV   γ   4038

γ

4361   R   4027

λ   ψ

α

HYDRA

ζ   ε

CRATER

-20°

β   α

β

CRV
③

×1

Sky
Charts
2, 3, 8,

CRT
④

×1

Sky
Charts
2, 3, 8,
9

HYDRA

-30°

13ʰ   12ʰ   11ʰ   10ʰ

CENTAURUS

ANTLIA

-40°   Wil Tirion

Corvus & Crater

(KOR-vus) The Crow   (KRAY-ter) The Cup

The Crow and the Cup,
from the constellation cards
of Urania's Mirror (1825)

Arc to Arcturus, speed to Spica, then turn west and you will see a small foursome of stars that the ancients all called the Crow or the Raven. Crater is a fainter constellation alongside that looks like a cup.

Sent one day by Apollo for a cup of water, Corvus was slow in returning as he had been waiting for a fig near the spring to ripen. Bringing the cup (Crater) of spring water and a water serpent (Hydra) back in his claws, he told Apollo that he had been delayed because the serpent had attacked him. Apollo, knowing Corvus was lying, placed all three in the sky.

The Cup is to the west of Corvus, within reach, but the serpent prevents him from drinking from it.

NGC 4027 is a faint,
disturbed spiral galaxy
in Corvus.

R Corvi: This Mira–type variable star ranges from magnitude 6.7 to 14.4 over a period of about 10 months.

Tombaugh's Star: This very faint cataclysmic variable star, **TV Corvi**, was discovered as a nova by Clyde Tombaugh in 1931, while searching for planets. It was then forgotten until I chanced upon it while researching a biography of Tombaugh. After examining 360 photographs, I found it had repeated its outburst 10 times! In 1990, I observed an explosion of the star with a 16 inch (400 mm) telescope.

The Ring-tailed Galaxy: Also called the Antennae or Rat-tailed Galaxy, **NGC 4038** and **NGC 4039** form a faint, 11th magnitude pair of galaxies that are interacting or colliding. Needing an 8 inch (200 mm) telescope to see, it is still one of the brightest pairs of connected galaxies.

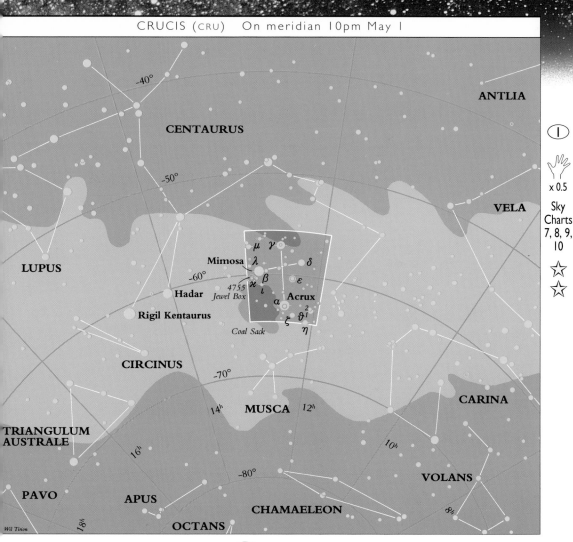

ANTLIA

CENTAURUS

-40°

-50°

VELA

LUPUS

μ γ

Mimosa λ

-60°   β

4755   ϰ   ι

Hadar   Jewel Box   ε

δ

Acrux

α

2
ϑ¹

ζ

η

Rigil Kentaurus

Coal Sack

CIRCINUS

-70°

CARINA

14ʰ   MUSCA   12ʰ

TRIANGULUM
AUSTRALE

16ʰ

10ʰ

-80°

VOLANS

PAVO   APUS

8ʰ

CHAMAELEON

OCTANS

Wil Tirion

18ʰ

①

🖐 x 0.5

Sky
Charts
7, 8, 9,
10

☆
☆

Crux

(KRUKS) The Southern Cross

The most famous southern constellation, the Southern Cross appears on the flags of several nations. Its distinctive pattern of stars helped guide sailors for centuries, the upright of the cross pointing the way to the south celestial pole. Because it lies so far south, Crux was not mapped as a separate entity until 1592. Until then, it formed part of Centaurus.

The cross contains the most striking pair of opposites—the Jewel Box and the Coal Sack—embedded within the southern Milky Way.

Acrux: This is the popular name for **Alpha (α) Crucis**, the bright double star at the foot of the cross, separated by about 4½ arc seconds. A third star, quite bright at magnitude 5, lies 90 arc seconds away.

Gamma (γ) Crucis: Also known as **Gacrux**, this wide double star marks the northern end of the cross. An optical double, it consists of a magnitude 6.4 star lying almost 2 arc minutes from a bright orange primary.

The Jewel Box: Superimposed on **Kappa (κ) Crucis**, this is one of the finest open

One feature of the Jewel Box is the contrasting colors of the line of three stars at its heart.

clusters. Although small, it sparkles in any instrument, and has several stars of contrasting color.

The Coal Sack: This is one of the largest and densest dark nebulae in the sky. It lies just east of Acrux and is clearly visible in a dark sky against the star clouds of the Milky Way.

163

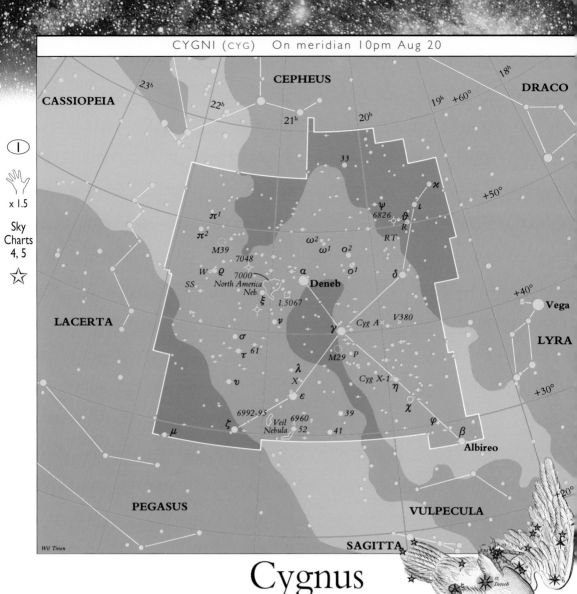

CASSIOPEIA

CEPHEUS

DRACO

LACERTA

Deneb

Vega

LYRA

PEGASUS

VULPECULA

SAGITTA

Wil Tirion

Cygnus

(SIG-nus) The Swan

Urania's Mirror (1825) places brilliant Deneb between the feet of the swan.

C ygnus is the Northern Hemisphere's answer to Crux. Looking like a large cross, Cygnus straddles the northern Milky Way, which is at its best in this part of the sky. If you are under a dark sky you may be able to see the Milky Way divide into two streams in Cygnus. A dark nebula between us and the more distant stars causes this apparent divergence.

Since the time of the Chaldeans, many civilizations have seen this constellation as a bird of some sort. One story claims that Cygnus is Orpheus, the great hero of Thrace, who sang and played his lyre so beautifully that wild animals and even the trees would come to hear him. It is said that Orpheus was transported to the sky as a swan, so that he could be near his cherished lyre. Another myth claims that Cygnus is Zeus in the disguise of a swan—the form he took to seduce Leda of Sparta.

👁 **Deneb (Alpha [α] Cygni)**: Deneb means "tail" in Arabic, which is where this star is positioned on the swan. On a par with Rigel in Orion, it is one of the mightiest stars known—

25 times more massive and 60,000 times more luminous than the Sun. About 1,500 light years away, Deneb is by far the most distant star of the famous Summer Triangle, which it forms with Vega and Altair. Vega is 25 light years away and Altair only 16.

👁 **Albireo (Beta [β] Cygni)**: Whether you are observing this star from the dark of the country or from the middle of a city, Albireo, at the foot of the cross, is one of the prettiest sights in the sky. Without a telescope it is seen as a single star; a telescope transforms it into a spectacular double with a separation of 34 arc seconds. One member is golden yellow with a magnitude of 3, and the other is blueish with a magnitude of 5.

(Left) The Northern Cross in Cygnus lies astride the bright northern Milky Way. The North America Nebula (NGC 7000), visible just to the left of Deneb, is clearer in the image below.

61 Cygni: Dubbed the Flying Star because of its rapid motion relative to more distant stars, this double is easily separated in small telescopes. The two components revolve around each other over the course of about 650 years. 61 Cygni seems to have one or more unseen companions, objects usually estimated to be 5 to 10 times the mass of Jupiter. If real, these would be large planets, but too small to be true stars.

The North America Nebula (NGC 7000): One of the sky's best examples of a bright nebula, this giant cloud is illuminated by Deneb, which lies only 3 degrees to the west. Because of its size, the nebula is difficult to see in a telescope: it is best seen with the naked eye on

a dark night. Photographs show this nebula to look surprisingly like the shape of North America, but this resemblance is not readily apparent to the eye when observing.

M 39: This loosely bound open star cluster is seen at its best through a pair of binoculars. On a clear night you might be able to see it with the naked eye, as Aristotle apparently did in around 325 BC.

Chi (χ) Cygni: At maximum brightness, typically magnitude 4 or 5, this long-period variable is bright enough to be seen with the naked eye. It fades to about magnitude 13 and then climbs back in a period of a little more than 13 months.

SS Cygni: One of numerous faint variables in Cygnus, SS Cygni is a dramatic cataclysmic variable star, erupting every two months to magnitude 8 but normally remaining faint at magnitude 12.

The Veil Nebula (NGC 6960, 6992, 6995): The lacy remnants of an ancient supernova, this beautiful nebulosity requires at least a 6 inch (150 mm) telescope. NGC 6960, the nebula's western arc, passes through 52 Cygni, which makes it easier to find but harder to see.

The Blinking Nebula (NGC 6826): This planetary nebula has a relatively bright central star. If you concentrate on the star, the surrounding cloud disappears.

NGC 6992 and 6995 form the eastern arc of the Veil Nebula.

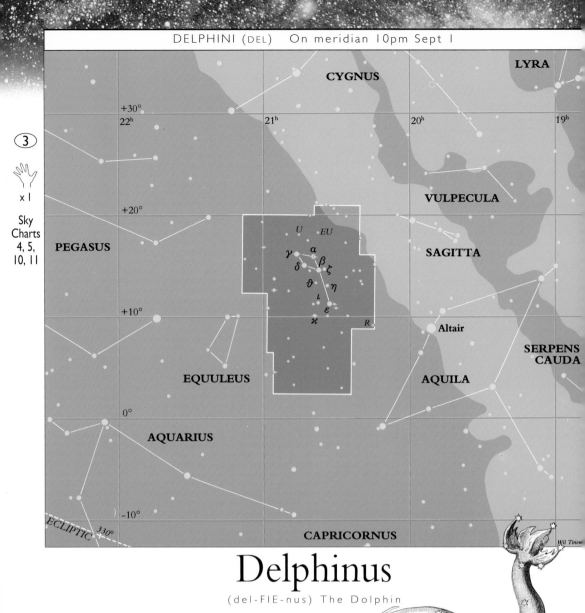

LYRA

CYGNUS

+30°
22ʰ   21ʰ   20ʰ   19ʰ

③

🖐
x 1

Sky
Charts
4, 5,
10, 11

VULPECULA

+20°

PEGASUS

U   EU

γ   α
δ   β ζ
ϑ   η
ι
κ   ε   R

+10°

SAGITTA

Altair

SERPENS
CAUDA

EQUULEUS

AQUILA

0°

AQUARIUS

-10°

ECLIPTIC   330°

CAPRICORNUS

Wil Tirion

Delphinus

(del-FIE-nus) The Dolphin

A small constellation with a distinctive shape, Delphinus has been thought of as a dolphin since ancient times. It is said that the mermaid Amphitrite agreed to marry Poseidon, from whom she had been trying to escape, on the advice of a dolphin. Poseidon was so pleased with the little dolphin that he placed him among the stars. The constellation is also sometimes known as Job's Coffin, the origin of which is obscure.

This small group of faint stars looks a little like a kite. Its alpha (α) star is named Sualocin and its beta (β) star is known as Rotanev. These names honor a relatively recent

Delphinus doesn't look quite like a modern dolphin in this engraving by Jacob de Gheyn (1621).

observer, Niccolo Cacciatore, long-time associate of the famous nineteenth-century observer Giuseppe Piazzi. Star atlases at the time included these names without comment, but the Reverend Thomas Webb worked out that the names, spelled backward, are Nicolaus Venator—the Latinized version of Cacciatore's name.

🔭 **Gamma (γ) Delphini**: This is an optical double with a separation of 10 arc seconds. The brighter is magnitude 4.5 and the fainter, which is slightly green, is 5.5.

🔭 **R Delphini**: This Mira star has a magnitude range of 8.3 to 13.3 over a period of 285 days.

NGC 6934 is a 9th magnitude globular, marked by the relatively bright star nearby.

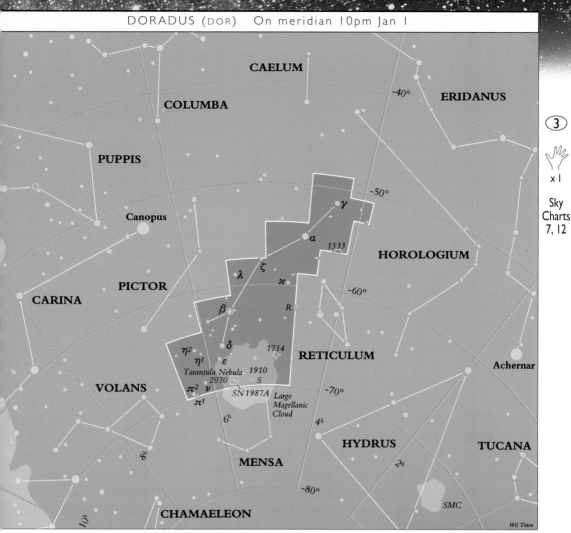

CAELUM

COLUMBA

ERIDANUS

~40°

PUPPIS

~50°

Canopus

γ

α

1533

HOROLOGIUM

λ   ζ

ϰ

PICTOR

~60°

CARINA

β

R

η²

η¹   ε

1714

RETICULUM

Achernar

Tarantula Nebula   1910

2070   S

VOLANS

π²   ν   SN 1987A   Large

π¹   Magellanic

Cloud

6ʰ   4ʰ

~70°

δ

HYDRUS

2ʰ

TUCANA

MENSA

~80°

SMC

CHAMAELEON

10ʰ

Wil Tirion

Sky
Charts
7, 12

③

x 1

Dorado

(doh-RAH-doh) The Goldfish, The Dolphinfish

Far to the south, this constellation was first recorded by Bayer in his star atlas of 1603. Dorado does not honor the tiny fish in many a home aquarium, but the tropical dolphinfish, the mahi-mahi, member of the Coryphaenidae family, which can be more than 5 ft (1.75 m) long. Since they swim fast and often leap out of the water in play, sailors used to consider their appearance a good omen.

👁 **The Large Magellanic Cloud (LMC)**: This is a companion galaxy to the Milky Way, lying 168,000 light years away—less than one-tenth the distance to the Andromeda Galaxy (M 31). As a result, it

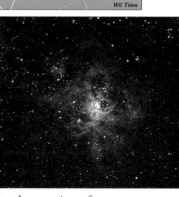

A cluster of hot, young stars centered around 30 Doradus illuminates the massive Tarantula Nebula.

spans about 11 degrees of the sky, presenting its contents to the scrutiny of Southern Hemisphere observers. It was from this galaxy that supernova 1987A blazed forth. The LMC is plainly visible in a dark sky, but it is easily lost in the glare of city lights.

The Tarantula Nebula (NGC 2070): Also known as the 30 Doradus Nebula, this is one of the finest emission nebulae in the sky, despite its distance from us. It is perhaps 30 times the size of the more famous Great Nebula in Orion (M 42).

S Doradus: A single super-luminous star within the open cluster **NGC 1910**, S Doradus varies irregularly in brightness between magnitudes 8 and 11. It is one of the most luminous stars known.

The Tarantula Nebula is the largest of the numerous pink nebulae in the Large Magellanic Cloud.

CAMELOPARDALIS

Polaris
+90°

URSA MINOR

CEPHEUS

+80°

URSA
MAJOR

CYGNUS

+70°

+60°

+50°

LYRA HERCULES BOÖTES

Wil Tirion

Draco

(DRAY-koh) The Dragon

This constellation is circumpolar from much of the Northern Hemisphere and is best seen during the warmer months. A large, faint constellation, the Dragon is hard to trace as it winds about between Ursa Major, Boötes, Hercules, Lyra, Cygnus, and Cepheus.

The Chaldeans, Greeks, and Romans all saw a dragon here, while Hindu mythology claims the creature is an alligator. The Persians saw a man-eating serpent.

Draco has been identified with a number of ancient Greek stories. A dragon guarded the entrance to the Hesperides, where the golden apples grew, and was killed by Hercules. And Athena threw a dragon into the sky, after it attacked her while she was fighting the Titans.

Thuban, the brightest star in the constellation, was the pole star in ancient times, but the Earth's precession has since moved the pole to Polaris.

👁 **Quadrantids:** This is one of the strongest meteor showers. The time of maximum activity is around January 3, and it lasts only a few hours.

👁 **Draconids:** This meteor shower consists of particles from Periodic Comet Giacobini-Zinner. In 1933 and 1946, the shower's date of October 9 closely followed the comet's crossing of the Earth's orbit, and the result was a storm of meteors.

🔭 **NGC 6543**: This 8th magnitude planetary nebula lies midway between the stars **Delta** (δ) and **Zeta** (ζ) **Draconis**. It is bright blue-green in color, but high power is needed in order to make out its small, hazy disk.

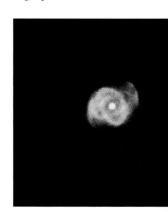

(Above) Draco, the Dragon, as portrayed by Johann Bayer in his Uranometria *(1603), curls its way through the heavens.*
(Right) NGC 6543 is one of the brighter planetary nebulae.

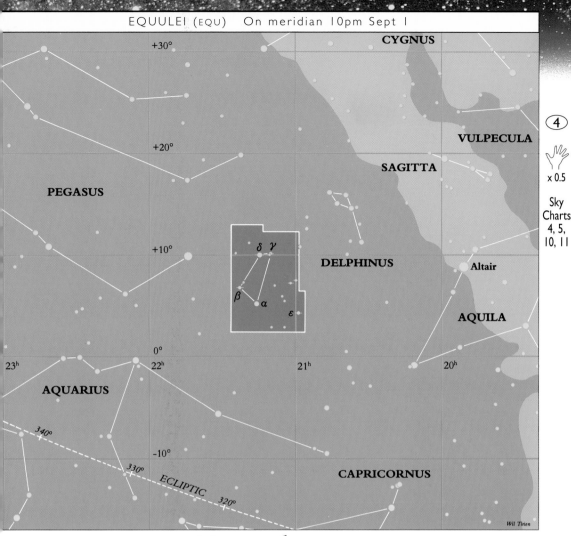

CYGNUS

+30°

VULPECULA

+20°

SAGITTA

PEGASUS

DELPHINUS

+10°

δ γ

Altair

β

α

ε

AQUILA

0°

23ʰ   22ʰ   21ʰ   20ʰ

AQUARIUS

340°

-10°

330°

ECLIPTIC   320°

CAPRICORNUS

④

× 0.5

Sky
Charts
4, 5,
10, 11

Wil Tirion

Equuleus

(eh-KWOO-lee-us) The Little Horse

With the exception of Crux, Equuleus occupies a smaller space than any other constellation. It lies just to the southeast of Delphinus and because it has no bright stars it is of limited interest. **Alpha (α) Equulei**, its brightest star, is named Kitalpha—Arabic for "little horse".

The famous Greek astronomer Hipparchus is thought to have made up the constellation in the second century BC. It has been said to represent Celeris, brother of Pegasus (the Winged Horse), given to Castor (one of the twins represented by Gemini) by Mercury.

Equuleus, as represented in Bayer's Uranometria (1603). It is likely that it came to be known as the Little Horse to distinguish it from nearby Pegasus, the Winged Horse.

SKYWATCHING TIP

When skywatching, it's a good idea to keep a record of what you see. Not only will this help you to remember your observations, but the effort to record details will stimulate careful and meaningful viewing. A notebook that's large enough for drawings as well as notes is ideal. The information you take down should include: date, time, and location of observation; instrument(s) used; seeing conditions; and sketches of your sightings.

Nebulae, star clusters, and galaxies present some of the most interesting targets for amateur astronomers. In general, low magnification and a wide field of view are necessary for deep-sky observing, but don't be afraid to experiment.

169

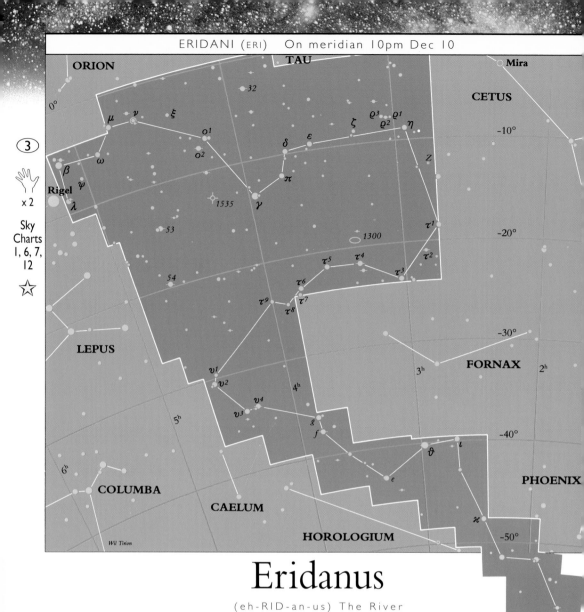

ORION   TAU   Mira

CETUS

ORION

Rigel

x 2

3

Sky
Charts
1, 6, 7,
12

☆

LEPUS

FORNAX

COLUMBA

CAELUM

HOROLOGIUM

PHOENIX

Wil Tirion

Eridanus

(eh-RID-an-us) The River

What a long constellation the river Eridanus is! Its source lies immediately to the west of Rigel in Orion, with a star called Cursa or Beta (β) Eridani. It flows southward until it reaches its mouth in **Achernar (Alpha [α] Eridani)** near the south celestial pole—a very bright star few northern observers ever see. From the Southern Hemisphere, an observer can follow the full course of the river, even though the stars are faint.

This constellation has been seen as a river since ancient times—usually the Euphrates or the Nile. For early observers in South-west Asia, the river extended only as far south as Acamar, or **Theta (θ) Eridani**, because

Really a target only for larger telescopes, NGC1300 is a classic barred spiral.

they could not see the stars that lay further south. In Book II of *Metamorphoses*, Ovid writes of Phaethon being tossed out of the chariot of the Sun to drown in Eridanus.

Omicron 2 (o₂) Eridani: This remarkable triple consists of a 4th magnitude orange dwarf, a 9th magnitude white dwarf, and an 11th magnitude red dwarf. The red and white dwarfs form a pair (separation 8 arc seconds), and are separated from the brighter star by over 80 arc seconds. The white dwarf is the only one of its class that is easy to see in a small telescope.

Epsilon (ε) Eridani: Only 10.8 light years away, this star is a smaller version of our Sun. Radio telescopes have been pointed towards it in a so far unsuccessful search for signals indicating intelligent life.

Eridanus, the River, with brilliant Achenar at its mouth, in Bayer's drawing of 1603

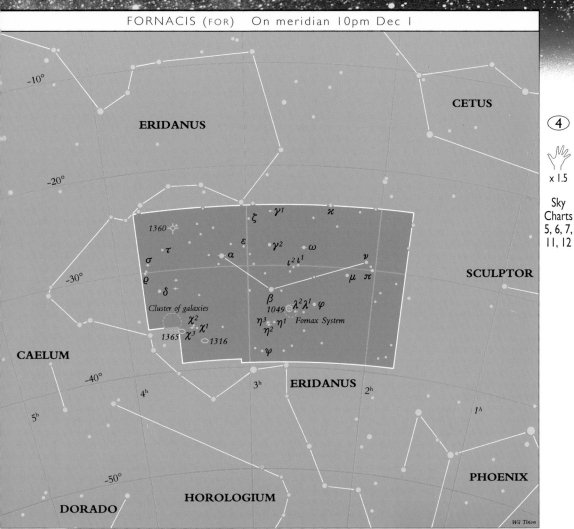

-10°

-20°

-30°

-40°

-50°

ERIDANUS

CETUS

1360

ζ   γ¹   κ

τ   ε   γ²   ω

σ   α   ι²ι¹   ν

ρ   μ π

δ

Cluster of galaxies   β   λ²λ¹ φ
1049   η³   η¹   Fornax System
χ²   η²
1365   χ³ χ¹   1316
ψ

SCULPTOR

CAELUM

4ʰ   3ʰ   ERIDANUS   2ʰ

5ʰ   1ʰ

-50°

PHOENIX

HOROLOGIUM

DORADO

Wil Tirion

4

×1.5

Sky
Charts
5, 6, 7,
11, 12

Fornax

(FOR-nax) The Furnace

NGC 1365, a barred spiral
galaxy, and one of the
brighter Fornax galaxies, at
9th magnitude

When Nicolas-Louis de Lacaille invented this constellation out of several faint stars in a bend of River Eridanus, he was honoring the famous French chemist Antoine Lavoisier, who was guillotined during the French Revolution in 1794.

The Fornax Galaxy Cluster: While there are no bright points of interest in Fornax, if you have a large telescope you will enjoy this challenging cluster of galaxies near the Fornax–Eridanus border. With a wide-field eyepiece, you may see up to nine galaxies in a single field of view. The brightest galaxy, at 9th magnitude, **NGC 1316** is also the radio source Fornax A.

The Fornax System: Although this dwarf galaxy—a diminutive member of our Local Group of galaxies—appears to be unusual, galaxies like it may be common in the universe. Spherical in shape, it is a large group of very faint stars that includes a few globular clusters. It is too faint to see with an amateur telescope, but one globular cluster, **NGC 1049**, is, at magnitude 12.9, visible in a 10 inch (250 mm) telescope under a good sky.

A portion of the Fornax Galaxy Cluster, in a field about
1 degree across

171

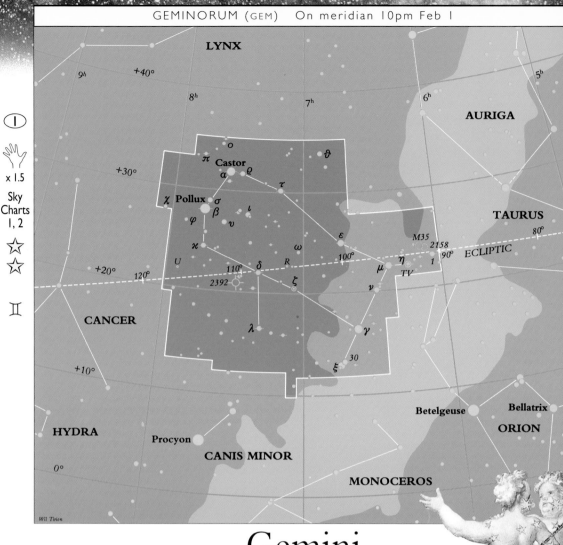

LYNX

AURIGA

TAURUS

ECLIPTIC

CANCER

HYDRA

Procyon

CANIS MINOR

MONOCEROS

Betelgeuse

Bellatrix

ORION

Wil Tirion

x 1.5
Sky
Charts
1, 2

Gemini

(JEM-eh-nye) The Twins

A familiar pattern in the sky, Gemini is part of the zodiac. Various cultures have seen the stars as twins—either as gods, men, animals, or plants. The Greeks named the constellation's two brightest stars Castor and Pollux, after the twins who hatched from an egg from their mother Leda, following her seduction by Zeus. The twins were among the heroes who sailed with Jason in the quest for the Golden Fleece. They helped save the *Argo* from sinking during a storm, so the constellation was much valued by sailors.

William Herschel discovered Uranus near Eta (η) Geminorum in 1781, and Clyde Tombaugh discovered Pluto near Delta (δ) Geminorum in 1930.

M 35, with NGC 2158 just to the west

Gemini, as depicted in a fresco from an Italian villa (1575)

Castor (Alpha [α] Geminorum):
This sextuple star can be seen only as a double through a small telescope. Its current separation is about 3 arc seconds.

Eta (η) Geminorum: This bright semi-regular variable varies from magnitude 3.2 to 3.9 and back over about eight months.

M 35: This bright open cluster is beautiful through binoculars and spectacular in a small telescope. **NGC 2158** is a smaller, fainter open cluster on its southwest edge. It appears in small telescopes as a smudge, being about 16,000 light years away—five times the distance to M 35.

The Clownface or Eskimo Nebula (NGC 2392): This strange-looking, 8th magnitude planetary nebula has a bright central star. The blue-green tint of its 40 arc second disk gives it away.

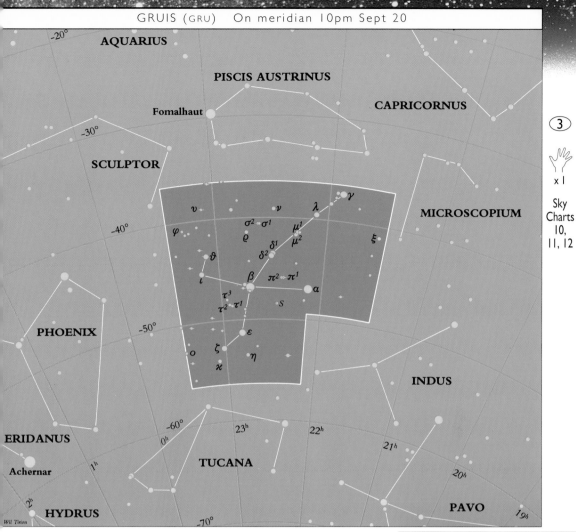

-20°
AQUARIUS
PISCIS AUSTRINUS
CAPRICORNUS
Fomalhaut
-30°
SCULPTOR
MICROSCOPIUM
-40°
ν   ν   λ   γ
φ   σ² σ¹
μ¹
ϱ   δ¹ μ²   ξ
ϑ   δ²
ι   β   π² π¹   α
τ³
τ² τ¹   S
PHOENIX   -50°
ε
o   ζ   η
ϰ
INDUS
-60°
ERIDANUS   0ʰ   23ʰ   22ʰ
21ʰ
Achernar   1ʰ   TUCANA
20ʰ
2ʰ   PAVO   19ʰ
HYDRUS   -70°
Wil Tirion

③
🖐
x 1
Sky
Charts
10,
11, 12

Grus

(GROOS) The Crane

I n his star atlas of 1603, Johann Bayer named this southern constellation Grus, the Crane— the bird which served as the symbol of astronomers in ancient Egypt. This group of stars—variously seen as a stork, a flamingo, and a fishing rod—has very little to offer the skywatcher who is using a small telescope, although some faint galaxies provide suitable targets for telescopes of 8 inch (200 mm) aperture or larger.

It is interesting to compare the Crane depicted by Bode in Uranographia (1801) with our chart of Grus.

The cross-like shape of Grus is easily found just south of bright Fomalhaut

Grus has only three fairly bright stars, which can be used as a simple illustration of magnitude.
   Alpha (α) Gruis, also known as Alnair, is a large, blue main-sequence star about 70 times as luminous as the Sun. Being only 57 light years away, it is the brightest of the three stars only because it is relatively close to us.
   Beta (β) Gruis is a much larger red giant star, some 800 times as luminous as the Sun, but its 140 light year distance from us makes it appear fainter than Alpha (α) Gruis.
   Finally, **Gamma (γ) Gruis**, a blue giant star, which is actually more luminous than either of the others, appears fainter than them since it is 230 light years away.

173

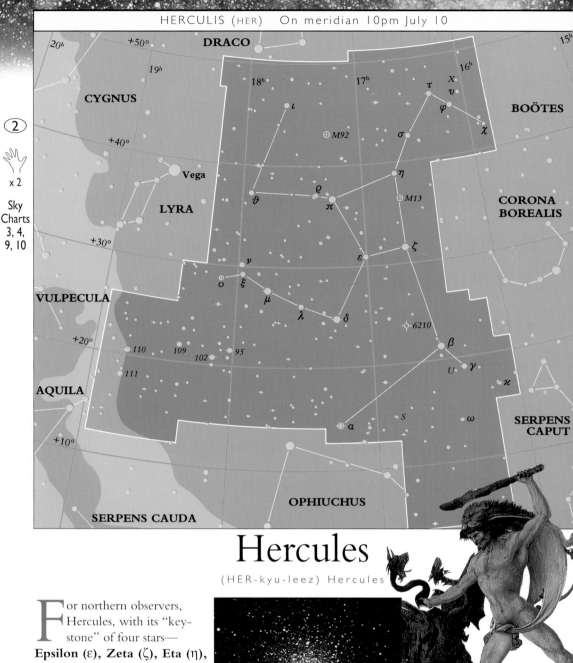

2

🖐 x 2

Sky Charts 3, 4, 9, 10

20ʰ   +50°   DRACO   15ʰ

19ʰ   18ʰ   17ʰ   τ   X   16ʰ

CYGNUS   υ   φ   BOÖTES

+40°   ⊕ M92   σ   χ

Vega   η

LYRA   ϑ   ϱ   π   ⊕ M13   CORONA BOREALIS

+30°   ζ

ν   ε

VULPECULA   ο   ξ   μ

+20°   λ   δ   6210

110   109   95   β

111   102   U   γ   κ

AQUILA   ω   SERPENS CAPUT

+10°   S   α

OPHIUCHUS

SERPENS CAUDA

Hercules

(HER-kyu-leez) Hercules

F or northern observers, Hercules, with its "key-stone" of four stars— **Epsilon (ε), Zeta (ζ), Eta (η), and Pi (π)**—is one of the best of the summer constellations.

One of the most famous of all the classical heroes, Hercules was immensely strong and was revered throughout the Mediterranean. He was the half-mortal son of Jupiter and was involved in many noble exploits, the most famous being the undertaking of the twelve labors. At the end of his life, as a reward for his bravery, Jupiter made him one of the gods, placing him in the sky.

👁 **The Hercules Cluster (M 13):** The most dramatic globular cluster in the northern sky, this is faintly visible to the naked eye as a fuzzy spot, but through a telescope it is a sight to behold. The edges begin to resolve into stars in a 6 inch (150 mm) telescope. When you view this cluster, you are looking 23,000 years into the past.

(Left) M 13, the premier globular cluster of the northern sky. (Right) Hercules and the Hydra, from a painting by Antonio Pollaiuolo (1432–98)

M 92: M 13's slightly smaller and fainter cousin, this cluster of stars is some 26,000 light years away.

👁 **Ras Algethi (Alpha [α] Herculis):** This is a very red star, varying from magnitude 3.1 to 3.9. It is also a splendid colored double, with a 5th magnitude blue-green companion about 5 arc seconds away from an orange primary.

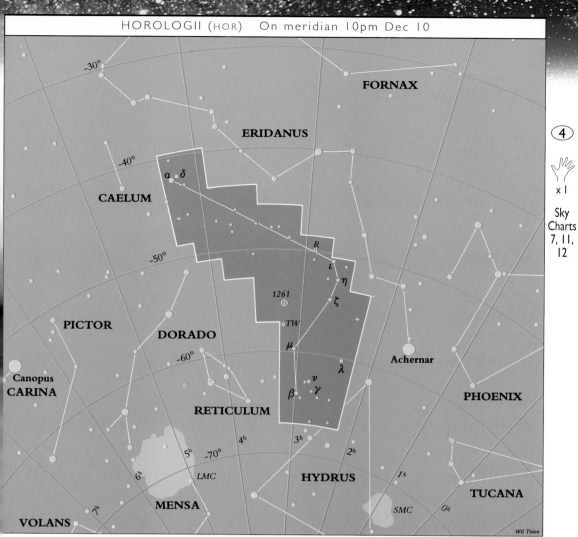

FORNAX

ERIDANUS

CAELUM

α δ

-30°

-40°

-50°

R

ι

η

1261

ζ

TW

μ

λ

ν

β γ

Achernar

PICTOR

DORADO

-60°

Canopus
CARINA

RETICULUM

4ʰ

3ʰ

2ʰ

PHOENIX

5ʰ   -70°

6ʰ

LMC

1ʰ

HYDRUS

SMC   0ʰ

TUCANA

VOLANS

MENSA

Wil Tirion

④

x 1

Sky
Charts
7, 11,
12

Horologium

(hor-oh-LOH-jee-um) The Clock

A small group of stars lying east of Archernar, this is one of the con–stellations mapped by Nicolas-Louis de Lacaille. Originally called Horologium Oscillatorium, it honors the invention of the pendulum clock by Dutch scientist Christiaan Huygens in 1656 or 1657.

A 10 inch (250 mm) telescope captured this view of NGC 1261, seen at a distance of 70,000 light years.

Huygens was a leader in Renaissance thinking. By apply-ing the law of the pendulum discovered by Galileo to clockmaking, he significantly increased the accuracy of timekeeping. Huygens' second great contribution to science was the discovery of Saturn's ring.

R Horologii: This long-period variable star was discovered from an observing station that Harvard University used to run in Peru. In 13½ months, it completes its cycle of variation from 5th to 14th magnitude and back.

NGC 1261: This 8th magnitude globular cluster is only 6 arc minutes across so is a target for a larger telescope.

SKYWATCHING TIP

The vigil of meteor observation can be a more pleasurable experience when done in groups. The entire sky can be observed continuously over a longer period, with each person allocated a particular segment to monitor, and there is the added benefit of conversation to keep participants from falling asleep! Since effective meteor watching entails constant surveillance of the sky, sleeping bags and reclining chairs are recommended to prevent the inevitable stiff neck from setting in. Comfort, as well as patience, are necessary for producing useful results.

+10°

0°

14ʰ

13ʰ

15ʰ

VIRGO

ECLIPTIC   180° 12ʰ

190°

200°

-10°

210°

Spica

CR.

220°

CORVUS

230°

-20°

R   γ   ψ

M68

LIBRA

π

M83

CENTAURUS

58   -30°

β

LUPUS

Wil Tirion

③
🖐
x 5

Sky
Charts
2, 7, 8,
9

Hydra

(HY-dra) The Sea Serpent

Hydra was the nine-headed serpent that Hercules had to kill as one of his twelve labors. Each time he lopped off one head, two others grew in its place. Hercules emerged from this nightmare by having his nephew burn the stump of each severed neck, preventing new heads from sprouting. In the midst of the struggle, Juno sent Cancer the crab to attack Hercules and distract him. The crab nipped Hercules, who then stepped on it and killed it. For its bravery, Juno rewarded the crab with a place in the sky (see Cancer p. 144).

As in the case of other large constellations, some mapmakers have tried to break up the snaking form of Hydra. In 1805, the French astronomer Joseph Lalande entertained himself by making up a constellation called Felis, the Cat. "I am very fond of cats," Lalande wrote.

"The starry sky has worried me quite enough in my life, so that now I can have my joke with it." Lalande formed his feline from stars of Hydra and Antlia, but it has not survived. Hydra remains, snaking a quarter of the way across the sky.

R Hydrae: One of the earliest known variables, this Mira star's light changes were first seen in the late 1600s. It varies over 13 months from a maximum as high as magnitude 3.5 to a minimum of 10.9.

V Hydrae: A rare example of a carbon star, this is a low-temperature red giant producing carbon. It is so deeply red that you can be sure you have found it merely by its color. The star varies somewhat erratically between magnitudes 6 and 12, with two superimposed periods—one about 18 months, and the other about 18 years.

In Uranometria, (1603), Bayer featured Hydra as a sea serpent rather than as the nine-headed serpent.

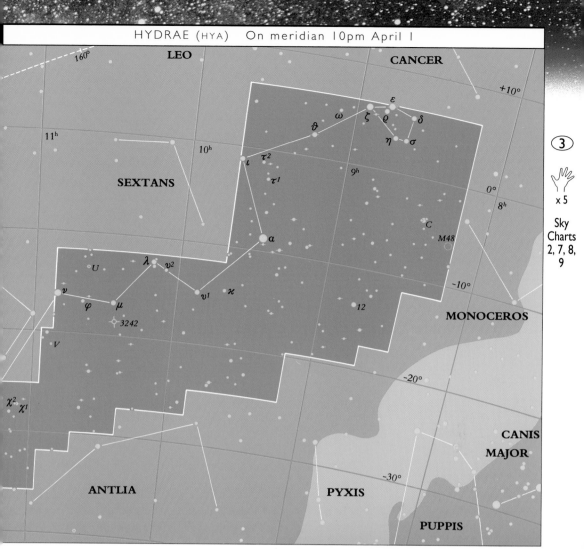

LEO

CANCER

+10°

160°

11ʰ

10ʰ

SEXTANS

ε

ω ζ ρ δ

ϑ

η σ

τ²

τ¹

ι

9ʰ

0°

8ʰ

α

C

M48

λ v²

U

v¹ κ

v

φ μ

3242

12

MONOCEROS

–10°

V

–20°

χ² χ¹

CANIS

MAJOR

–30°

ANTLIA

PYXIS

PUPPIS

③

✋ x 5

Sky
Charts
2, 7, 8,
9

M 48 (NGC 2548): Long considered a missing Messier object because he wrongly reported its position, M 48 is now considered to be one and the same as NGC 2548—a large open cluster best seen through binoculars or a wide-field telescope.

M 83: This is a strange-looking spiral galaxy with three obvious spiral arms. At 8th magnitude, it is one of the brighter galaxies visible in binoculars and will show more detail at higher magnification in a telescope. Watch out for "new" stars here, as M 83 has produced four supernovae in the last 60 years.

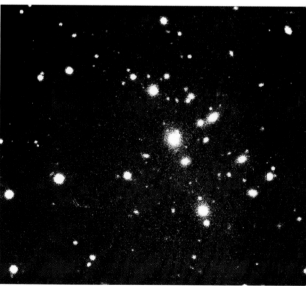

A CCD image of the distant Hydra cluster of galaxies. The brightest galaxies here are magnitude 18!

M83 is often classified as a barred spiral, and a bar across the nucleus is apparent in this amateur photograph.

The Ghost of Jupiter Nebula (NGC 3242): This nebula is the brightest planetary nebula in this part of the sky. It is about 16 arc seconds across and shows its structure well in 10 inch (250 mm) telescopes and larger.

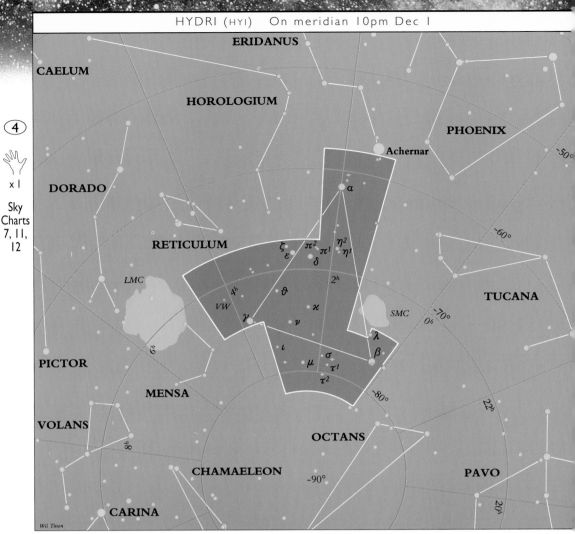

ERIDANUS

CAELUM

HOROLOGIUM

PHOENIX

Achernar

-50°

④

🖐
x 1

Sky
Charts
7, 11,
12

DORADO

RETICULUM

Achernar

α

-60°

ζ ε   π² π¹   η²
δ   η¹

LMC

2ʰ

TUCANA

4ʰ   ϑ

VW   κ

-70°

γ   ν   SMC   0ʰ

ι   λ

μ   σ   τ¹   β

τ²

-80°

PICTOR

MENSA

6ʰ

VOLANS

8ʰ

OCTANS

22ʰ

CHAMAELEON   -90°

PAVO

20ʰ

CARINA

Wil Tirion

Hydrus

(HY-drus) The Water Snake

Johann Bayer created this constellation,
publishing it in his star atlas of 1603. He
placed it near Achernar, the mouth of the
River Eridanus and nestled between the Large
and Small Magellanic Clouds. It is sometimes
called the Male Water Snake, to avoid confusing
it with Hydra.

🔭 **VW Hydri**: This star is the most popular
cataclysmic variable with Southern
Hemisphere observers. When in its usual state,
it shines at a faint 13th
magnitude, but when it
goes into outburst, an
event which occurs about
once a month, it can
become brighter than 8th
magnitude in just a few
hours.

*In his Uranographia (1801), Bode
shows Hydrus snaking past
Nubecula Minor, the Small
Magellanic Cloud in Tucana.*

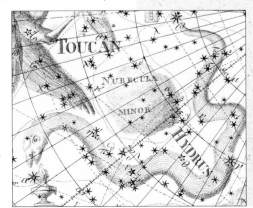

SKYWATCHING TIP

One frigid night in Flagstaff, Arizona, Clyde
Tombaugh—the astronomer who discovered
Pluto—was staring into the eyepiece of his
telescope, guiding a one-hour exposure. Feeling
sleepy, he struggled to keep from dozing off. When
the hour was up, he realized that he had become
so cold that he could hardly move. In great pain, he
managed to close the
telescope and move to a
warm room, where he had
to sit for some time beside
a heater to thaw out.

Be careful not to become
so absorbed in your
observations that you are
unaware of how cold you
are. Walk around the
telescope to stimulate your
circulation, or go inside.

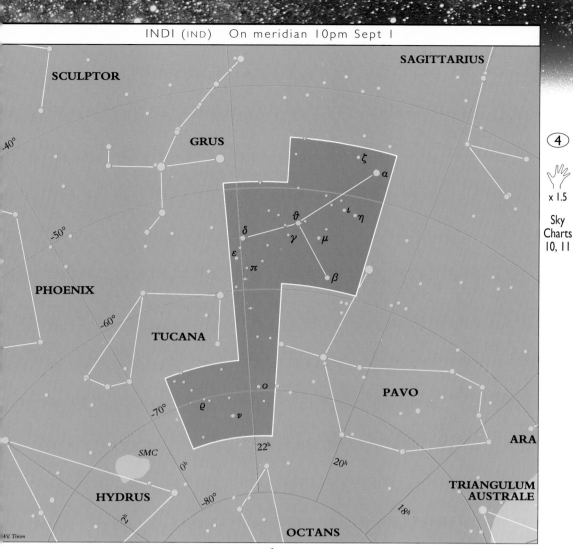

SCULPTOR

SAGITTARIUS

GRUS

ζ

α

-40°

ι η
ϑ
δ
γ μ
ε
π

β

PHOENIX

-50°

-60°

TUCANA

ο

PAVO

ARA

-70°

ϱ
ν

SMC
0ʰ

22ʰ

20ʰ

TRIANGULUM
AUSTRALE

HYDRUS

-80°

18ʰ

OCTANS

Wil Tirion

4

x 1.5

Sky
Charts
10, 11

Indus

(IN-dus) The Indian

T his constellation was added to the southern sky by Johann Bayer to honor the Native Americans that European explorers encountered on their travels. The figure of Indus is positioned between three birds: Grus, the Crane; Tucana, the Toucan; and Pavo, the Peacock.

Epsilon (ε) Indi: Only 11.3 light years away, this is one of the closest stars to the Sun and is somewhat similar to it.

With four-fifths of the Sun's diameter and one-eighth its luminosity, scientists consider Epsilon (ε) Indi to be worth investigating for planets and for evidence of extraterrestrial intelligence, such as radio signals.

In the early 1960s, when Frank Drake began searching for signs of life elsewhere in the galaxy, he used this star as one of his targets. In 1972, the Copernicus Satellite searched unsuccessfully for laser signals from this star.

*...dus, standing between two of its feathered neighbors:
...ucana (left) and Pavo (right)*

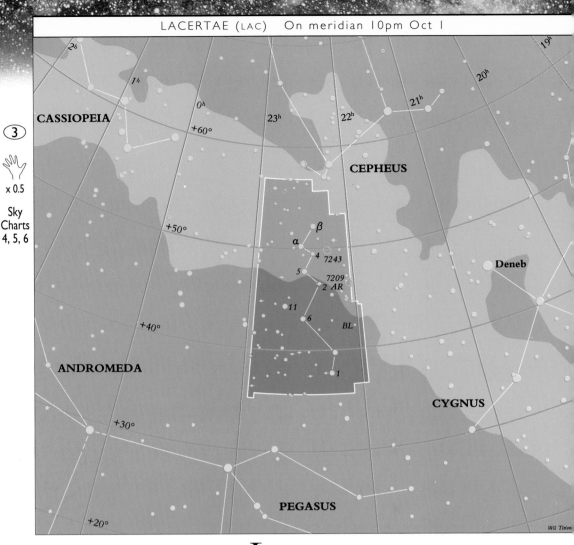

CASSIOPEIA

CEPHEUS

β

α

4 7243

5

7209
2 AR

11

6

BL

Deneb

ANDROMEDA

1

CYGNUS

PEGASUS

Wil Tirion

③
🖐
× 0.5

Sky
Charts
4, 5, 6

+60°
+50°
+40°
+30°
+20°

2ʰ 1ʰ 0ʰ 23ʰ 22ʰ 21ʰ 20ʰ 19ʰ

Lacerta

(lah-SIR-tah) The Lizard

Lacerta is far enough north to be circumpolar at the higher mid-northern latitudes. It lies south of Cepheus.

The German astronomer Johannes Hevelius suggested that this group of stars be named Lacerta in 1690, but a few revisions were needed before it evolved from a small, long-tailed mammal into a lizard. Other cartographers came up with names for the region to honor France's Louis XIV and Prussia's Frederick the Great, but these names were ignored.

BL Lacertae: Since this object varies from 13.0 to 16.1, it is invisible to any but the largest amateur telescopes. However, BL Lacertae is worth a look, since it is not a star at all but the nucleus of a distant elliptical galaxy. Some of this class of BL Lacertae-type (BL Lac) objects have been known to change by as much as two magnitudes in a single day. Recent theories suggest that

BL Lac objects, quasars, and other high-powered galaxies are all closely related "active galaxies". This powerful energy source at the center may be a black hole surrounded by a complex, swirling mass of gas and dust.

SKYWATCHING TIP

As you become more involved in observing and develop connections with your local astronomy club, you might be invited to a star party sponsored by the club. Take full advantage of this wonderful opportunity to meet people, ask questions, share experiences, compare notes, and look through the various telescopes that are being used. Remember to take a red flashlight with you, so that you can operate effectively in the dark without inconveniencing anyone, and always point it downward, away from people's dark adapted eyes.

Lacerta, the Lizard, as depicted in the constellation cards Urania's Mirror *(1825)*

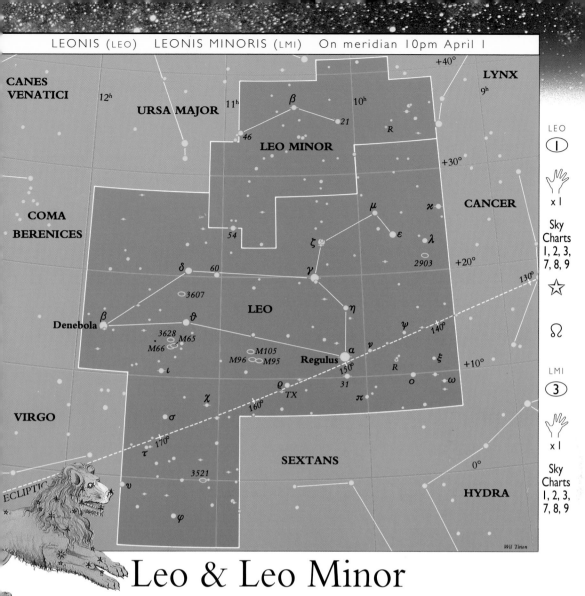

LEO

①

x 1

Sky
Charts
1, 2, 3,
7, 8, 9

☆

♌

LMI

③

x 1

Sky
Charts
1, 2, 3,
7, 8, 9

Leo & Leo Minor

(LEE-oh) The Lion   (LEE-oh MY-ner) The Little Lion

Unlike most of the zodiacal constellations, Leo, with its sickle (or backward question mark) tracing out a great head, really can be pictured as its namesake, a lion reclining not unlike the Egyptian Sphinx.

The Babylonians and other cultures of Southwest Asia associated Leo with the Sun, because the summer solstice occurred when the Sun was in that part of the sky.

Leo Minor is a recent addition to the constellations, introduced by Johannes Hevelius during the seventeenth century.

Gamma (γ) Leonis: This beautiful double star has orange–yellow components of 2nd and 3rd magnitude separated by 5 arc seconds.

R Leonis: This favorite Mira variable is easily found near **Regulus**. It ranges from magnitude 5.9 to 11 over about 10½ months.

R Leonis Minoris: Another Mira star, taking about a year to vary between magnitudes 7.1 and 12.6.

M 65 and M 66: These two spiral galaxies near **Theta (θ) Leonis** are visible in binoculars but give a better view in a telescope. Other interesting galaxies in Leo are **NGC 3628, M 95, M 96, M 105**, and **NGC 2903**.

Leonids: This meteor shower peaks annually on November 17. In 1966, observers recorded up to 40 meteors per second at its peak.

(Above, left) The Urania's Mirror (1825) Leo is leaping rather than reclining. (Below) This view of Leo is dominated by blue-white Regulus and orange Gamma (γ) Leonis in the sickle.

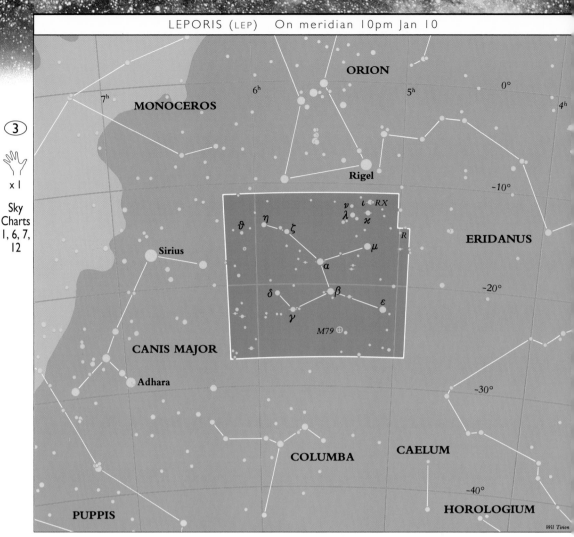

ORION

MONOCEROS

7ʰ   6ʰ   5ʰ   0°   4ʰ

Rigel

-10°

ν ι RX
λ ϰ
ϑ η ζ   R   ERIDANUS
μ

Sirius   α

δ   β   -20°
γ   ε

M79 ⊕

CANIS MAJOR

Adhara   -30°

COLUMBA   CAELUM

-40°

PUPPIS   HOROLOGIUM

Wil Tirion

③

🖐
x 1

Sky
Charts
1, 6, 7,
12

Lepus

(LEE-pus) The Hare

A faint constellation, Lepus is nevertheless easy to find because it is directly south of Orion. In ancient times it was thought of as Orion's chair. Egyptian observers saw it as the Boat of Osiris. The Greeks and Romans gave it the name Lepus. Since Orion particularly liked hunting hares, it was appropriate to place one below his feet in the sky.

🔭 **Gamma (γ) Leporis**: Easy to separate in virtually any telescope, this wide double star with contrasting colors has a separation of 96 arc seconds. It is relatively close to Earth, at a distance of 21 light years, and is part of the Ursa Major stream.

Hind's Crimson Star: Likened by some observers to a drop of blood in the sky, **R Leporis** is the variable that the nineteenth-century British

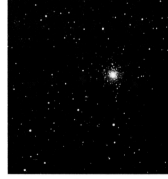

The globular cluster M 79, pictured using a 12 inch (300 mm) telescope

astronomer J. Russell Hind called the Crimson Star. Over a period of 14 months, the star varies in magnitude from a maximum of as much as 5.5 to a minimum of 11.7. Its coloring is at its most striking when the sky is dark and the star is near maximum brightness.

🔭 **M 79**: With the galactic center in Sagittarius half the sky away, this is a surprising place to find a globular cluster. Nevertheless, M 79 is here to enchant you, especially if you have an 8 inch (200 mm) or larger telescope which will begin to resolve the stars around its edges.

Lepus, from the constellation cards Urania's Mirror (1825)

182

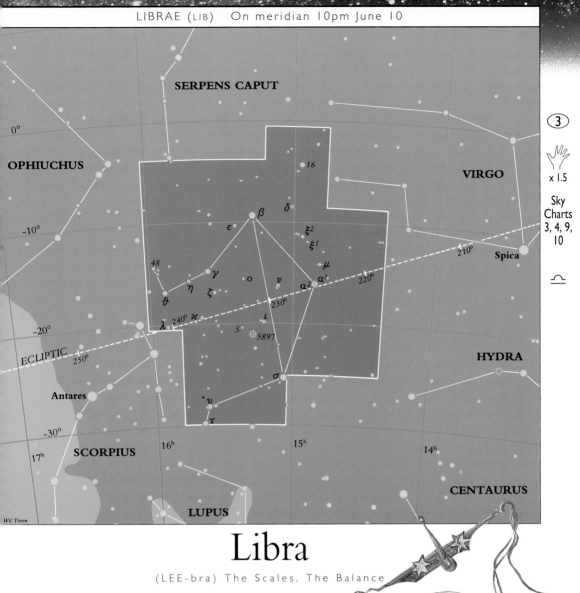

SERPENS CAPUT

0°

OPHIUCHUS

16

VIRGO

-10°

β

δ

ε

ξ²

ξ¹

210°

Spica

48

γ

μ

η   ζ

o

ν

α²   α¹

220°

ϑ

230°

λ   240° κ

ι

-20°

S

5897

HYDRA

ECLIPTIC

250°

Antares

υ

-30°

τ

SCORPIUS   16ʰ   15ʰ   14ʰ

17ʰ

CENTAURUS

Wil Tirion

LUPUS

③

🖐 x 1.5

Sky
Charts
3, 4, 9,
10

♎

Libra

(LEE-bra) The Scales, The Balance

*A detail from
a fresco at the Villa
Farnese, in Italy,
showing Libra, the Scales*

Looking like a high-flying kite, Libra is easy to find by extending a line westward from Antares and its two bright neighbors in Scorpius. The line reaches a point between Alpha (α) and Beta (β) Librae.

Libra is one of the constellations of the zodiac and was associated with Themis, the Greek goddess of justice, whose attribute was a pair of scales. Originally these stars were thought of as part of Scorpius: the alpha (α) and beta (β) stars both carry Arabic names, the former being Zuben El Genubi, "southern claw" and the latter Zuben

Libra leads the bright stars of Scorpius across the sky. The claws of the scorpion and brilliant red Antares dominate the lower corner of the image.

Eschamali, "northern claw". Our understanding is that Libra became a separate constellation at the time of the ancient Romans.

👁 **Delta (δ) Librae**: Similar to Algol, this eclipsing variable star fades by about a magnitude every 2.3 days, from 4.9 to 5.9. The entire cycle is visible to the naked eye.

🔭 **S Librae**: A Mira star, S Librae varies from an 8.4 maximum to a 12.0 minimum over a period of a little more than six months.

4

x 1

Sky
Charts
8. 9,
10

ECLIPTIC 250° 240°

Antares

-20°

-30°

ξ χ
ψ1
ψ2 φ1
ϑ   φ2
RU ⊕ 5986
η   ν
γ   δ
ω
ε   β ο   1.4406
λ   τ1
τ2
ν1 μ π   ι
ν2 κ   α
ϱ
σ
ζ

-40°

-50°

2

5822

16ʰ   15ʰ   14ʰ

17ʰ

Hadar   -60°   13ʰ
Rigil Kentaurus

18ʰ

Mimosa
12ʰ

Wil Tirion   Acrux

Lupus
(LOO-pus) The Wolf

South of Libra and east of Centaurus, Lupus the Wolf, is a small constellation with some 2nd magnitude stars. It is almost joined with Centaurus, as if the Centaur is stroking the wolf like a pet. The ancient Greeks and Romans called this group of stars Therion—an unspecified wild animal.

Lying within the band of the Milky Way, this constellation is home to a number of open and globular clusters.

SKYWATCHING TIP

Variable stars are named in a strange way. The first variable discovered in a constellation, say Lupus, is called R Lupi; the second S, then T, until Z. Then the names go RR Lupi, RS to RZ, then SS to SZ, and finally ZZ. In constellations with a great many variables, the list continues with AA to AZ, and on to QZ, but leaving out J. If more than these 334 variables are found, the system continues (more sensibly) with numbers, such as V 1500 Cygni for the 1975 nova in Cygnus.

🔭 **RU Lupi**: (Read this name out loud. After immersing yourself in constellation lore to this point, you possibly are!) RU Lupi is a faint nebular variable, with a maximum of only 9th magnitude. Its irregular variation is characteristic of young stars still involved with nebulosity.

NGC 5986, a globular that is visible in binoculars, close to some 6th and 7th magnitude stars

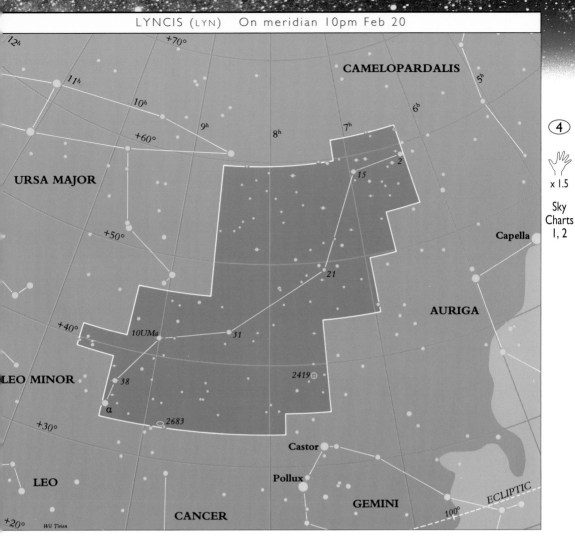

CAMELOPARDALIS

URSA MAJOR

Capella

AURIGA

LEO MINOR

α

2419

2683

LEO

Castor

Pollux

CANCER

GEMINI

ECLIPTIC

Wil Tirion

4

x 1.5

Sky
Charts
1, 2

Lynx

(LINKS) The Lynx

With only one 3rd magnitude star, Lynx is one of the hardest constellations to find. Johannes Hevelius charted this figure around 1690, apparently naming it Lynx because you need to have the eyes of a lynx to spot it. The same is true of its deep sky objects.

Almost edge-on, NGC 2683 is a 10th magnitude spiral galaxy just near the border with Cancer. Dust in the spiral arms is visible on the left side of the galaxy.

The Intergalactic Tramp (NGC 2419): Lying some 7 degrees north of Castor, the brightest star in Gemini, this is a very faint and distant globular cluster. It is more than 60 degrees from any

The lynx is a short-tailed cat of the Northern Hemisphere. The tail is rather long in this representation from Urania's Mirror (1825)

other globular. At 210,000 light years, it is more distant than the Large Magellanic Cloud and is so far away that it might escape the gravitational pull of our galaxy. It is for this reason that astronomer Harlow Shapley called it the Intergalactic Tramp. Through a 10 inch (250 mm) or larger telescope, NGC 2419 appears as a fuzzy knot of light.

185

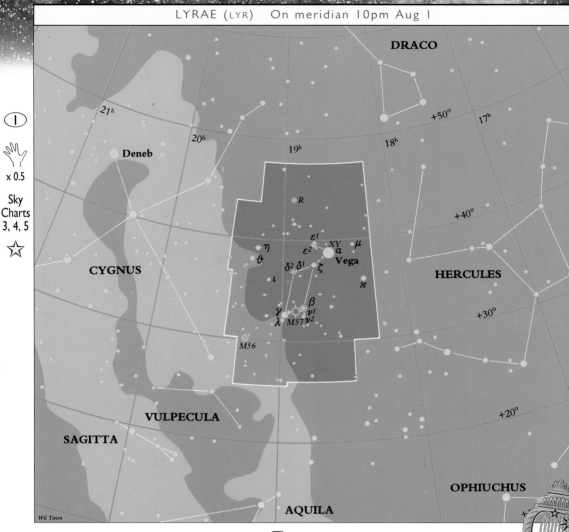

DRACO

+50°   17ʰ

21ʰ

20ʰ   19ʰ   18ʰ

Deneb

+40°

⊙R

ε¹   XY   μ
η   ε²   α
ϑ   δ² δ¹   ζ   Vega
ι   ϰ

CYGNUS   HERCULES

+30°

β
γ   ν¹
λ   M57 ν²
⊕
M56

+20°

VULPECULA

SAGITTA

OPHIUCHUS

AQUILA

Wil Tirion

I

🖐 x 0.5

Sky
Charts
3, 4, 5

☆

Lyra

(LYE-rah) The Lyre

T his beautiful constellation is dominated by Vega, one of the brightest stars in the sky. You can imagine the lyre strings stretched across the parallelogram of four stars that accompany it.

The lyre was given by Apollo to his son Orpheus. Orpheus played it so exquisitely that wild beasts and the mountains were enchanted. He passionately loved his wife Eurydice, and when she died he descended into the under-world to save her. He persuaded the gods to release her, on condition that he did not look at her on the journey. In his impatience, however, he glanced at Eurydice before they reached the upper world, and

The Lyre, with dazzling Vega near the top, from Urania's Mirror (1825)

she was swept back into Hades for ever. Inconsolable, Orpheus was torn to pieces by a group of young women after he ignored their advances. The lovers were then reunited and Orpheus's lyre was placed in the sky by Zeus.

Epsilon (ε) Lyrae: This is a "double double" star. The slightest optical aid shows two 5th magnitude stars—ε¹ and ε². Both are themselves doubles, with separations under 3 arc seconds. A 4 inch (100 mm) telescope operating at a magnification of 100 or more will split both of them.

Beta (β) Lyrae: This eclipsing variable ranges from magnitude 3.3 to 4.4 in 13 days.

The Ring Nebula (M 57): This famous planetary nebula lies midway between Beta (β) and **Gamma (γ) Lyrae**. Through a 3 inch (75 mm) or larger telescope it appears as a star out of focus at low magnification. Higher power will show its ring shape, about 2 arc minutes across.

A smoke ring in space? The famous Ring Nebula

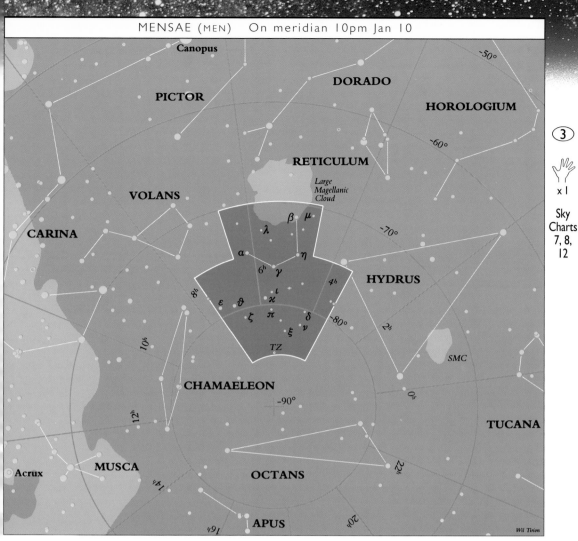

Canopus

PICTOR

DORADO

HOROLOGIUM

−50°

VOLANS

RETICULUM

Large Magellanic Cloud

−60°

CARINA

β μ

λ

α

η

−70°

6ʰ γ

HYDRUS

4ʰ

ε ϑ

ι

κ

ζ π δ ν

ξ

−80°

TZ

2ʰ

SMC

CHAMAELEON

−90°

0ʰ

TUCANA

Acrux

MUSCA

OCTANS

22ʰ

APUS

Wil Tirion

Mensa

(MEN-sah) The Table, The Table Mountain

The only constellation that refers to a specific piece of real estate, Mensa was originally called Mons Mensae by Nicolas-Louis de Lacaille, after Table Mountain, south of Cape Town, South Africa, where he did a good deal of his work. He developed this small constellation from stars between the Large Magellanic Cloud and Octans.

Mensa, without any bright stars, offers little to draw your attention away from the glories of the Large Magellanic Cloud. The northernmost stars of the constellation, representing the summit of the mountain, are hidden in the Large Magellanic Cloud, in the same way that Table Mountain is often shrouded in clouds.

👁 **Alpha (α) Mensae**: This dwarf star has an apparent magnitude of 5.1. Alpha (α) Mensae lies comparatively close to us, its light taking only 28 years to reach Earth.

👁 **Beta (β) Mensae**: Lying very near the edge of the Large Magellanic Cloud, this faint star of magnitude 5.3 lies at a distance of 155 light years away.

SKYWATCHING TIP

Almost all of the variable stars we mention can be seen with the naked eye at their brightest. However, our charts mainly include only the brighter stars, so if a variable drops below about 6th magnitude, as many do, it will be difficult to identify. Mira stars, though, can still be recognized by their reddish color.

Finding a variable star is only half the challenge. The rest involves estimating the magnitude of the star using nearby stars for comparison. For example, if Delta (δ) Cephei (see p.154) is a little fainter than the magnitude 3.5 star Zeta (ζ) and much brighter than the 4.4 Epsilon (ε), then Delta (δ) Cephei will be about magnitude 3.6 or 3.7.

But what if you do not know the magnitudes of the comparison stars, *a* and *b*? If the variable, *V*, were ¾ of the way from *a* to *b* in brightness, you could write your estimate as "*a*, 3, *V*, 1, *b*". This way you can monitor the changing brightness of *V* over time.

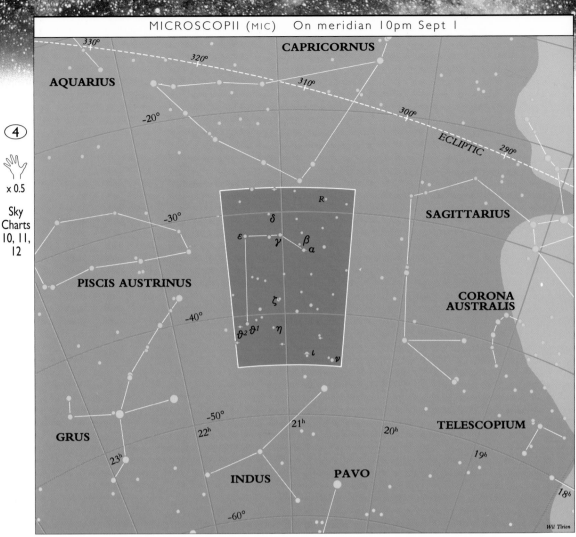

CAPRICORNUS

AQUARIUS

ECLIPTIC

SAGITTARIUS

PISCIS AUSTRINUS

CORONA
AUSTRALIS

GRUS

TELESCOPIUM

INDUS

PAVO

Wil Tirion

4

x 0.5

Sky
Charts
10, 11,
12

Microscopium

(my-kro-SKO-pee-um) The Microscope

This small, faint constellation, which lies just south of Capricornus and east of Sagittarius, was created by Nicolas-Louis de Lacaille in about 1750. It commemorates the microscope, the invention of which is credited to the Dutch spectacle-maker Zacharias Janssen, around 1590, and to Galileo, amongst others.

SKYWATCHING TIP

How good is your telescope? On a night when the atmosphere is steady, you can test its quality by trying to resolve double stars of equal magnitude. Beta (β) Monocerotis (opposite page), for example, is a triple system. Two stars, of magnitudes 4.7 and 5.2, form a pair separated by 2.8 arc seconds which should be split by a 2.4 inch (60 mm) telescope at 100x magnification. The third star, 7.3 arc seconds away, is even easier to split. Theta 2 (θ_2) Microscopii is another triple system but the close pair, separated by 0.5 arc seconds, is a test for a 10 inch (250 mm) telescope in good seeing.

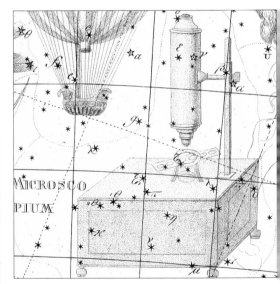

Bode depicted a microscope typical of his time (1801).

R Microscopii: This faint Mira variable has a rapid cycle lasting only 4½ months, during which it drops from magnitude 9.2 to 13.4 and climbs back again.

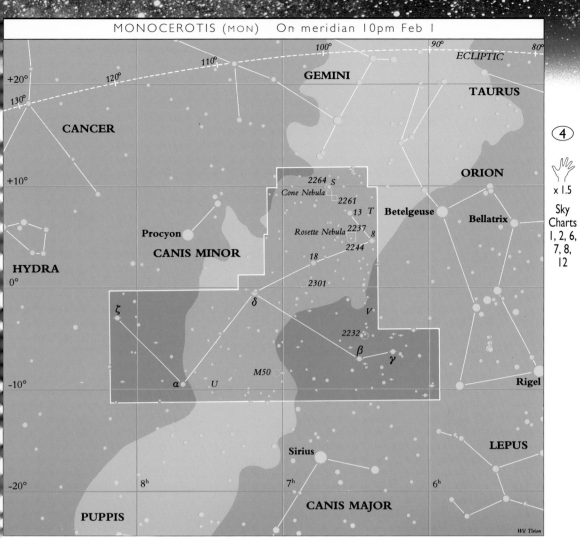

CANCER

GEMINI

TAURUS

ECLIPTIC

ORION

2264 S
Cone Nebula
2261
13 T　Betelgeuse
Rosette Nebula 2237　8
2244

Procyon

CANIS MINOR

Bellatrix

HYDRA

18

2301

ζ

δ

V

2232

β

γ

M50

α　U

Rigel

Sirius

LEPUS

PUPPIS

CANIS MAJOR

Wil Tirion

4

x 1.5

Sky
Charts
1, 2, 6,
7, 8,
12

Monoceros

(moh-NO-ser-us) The Unicorn

This faint constellation was formed in about 1624 by the German astronomer Jakob Bartsch. Monoceros is the Latin form of a Greek word meaning "one-horned", and it seems that the mythical unicorn may have come into existence as a result of a confused description of a rhinoceros.

Anxious to create a winter equivalent to the Northern Hemisphere's summer triangle, some observers advocate a winter triangle, bounded by Betelgeuse (in Orion), Sirius (in Canis Major), and Procyon (in Canis Minor). Monoceros, the Unicorn, and the band of the Milky Way fill the space inside this triangle.

The fabled unicorn, as seen in the constellation cards, Urania's Mirror *(1825)*

The blue-white stars of the Christmas Tree Cluster form a triangle which comes to a pinnacle at the top (south) of this picture.

M 50: This beautiful open cluster, lying slightly more than one-third of the way from Sirius to Procyon, is easy to find. Some of the cluster's stars are arranged in pretty arcs.

The Rosette Nebula (NGC 2237): Through a 10 inch (250 mm) telescope, this ring-shaped nebula, and the open cluster it contains (**NGC 2244**), offer a scene of delicate beauty. Smaller telescopes and binoculars will reveal the nebula on very clear nights.

The Christmas Tree Cluster (NGC 2264): This open cluster really does resemble a Christmas tree.

189

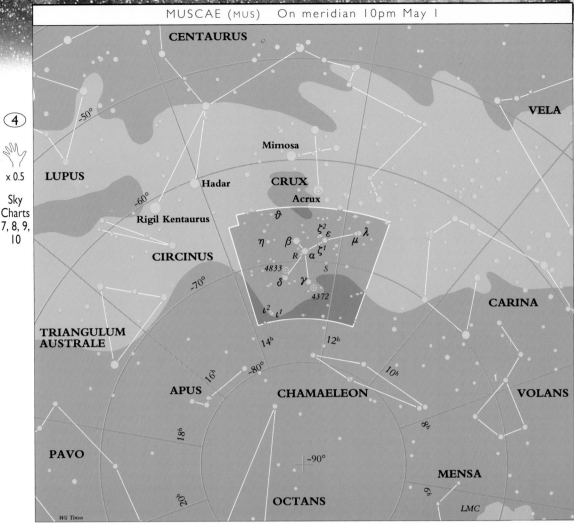

CENTAURUS

VELA

-50°

LUPUS

Mimosa

Hadar   CRUX
                Acrux

-60°

Rigil Kentaurus   ϑ

CIRCINUS   η   β   ζ²ε   μ   λ
                       R   α ζ¹
                  4833   S
-70°   δ   γ
                       4372

          ι²
          ι¹

TRIANGULUM
AUSTRALE

CARINA

14ʰ   12ʰ

16ʰ   -80°

APUS   CHAMAELEON   10ʰ   VOLANS

18ʰ   8ʰ

PAVO   MENSA

20ʰ   -90°   6ʰ

OCTANS   LMC

Wil Tirion

4

×0.5

Sky
Charts
7, 8, 9,
10

Musca

(MUSS-kah) The Fly

Musca is an easy constellation to find, just to the south of the Southern Cross. It was originally described by Johann Bayer in his 1603 star atlas as Apis, the Bee. Later on, Edmond Halley called it Musca Apis, the Fly Bee, then Nicolas-Louis de Lacaille named it Musca Australis, the Southern Fly—to avoid it being confused with the fly on the back of Aries, the Ram. Now that this northern fly is no longer a constellation, the Southern Fly is known simply as Musca.

Musca, the Fly, as featured in Johann Bode's Uranographia (1801)

The Southern Cross (Crux) and the Coal Sack overshadow the fainter stars of Musca, just to the south, on the edge of the glow of the Milky Way.

Beta (β) Muscae: This elegant double star consists of two 4th magnitude stars that revolve around each other in a period that spans several hundred years. The pair is some 520 light years from Earth. The separation of 1.6 arc seconds is very tight, presenting a challenge for a 4 inch (100 mm) telescope.

NGC 4372: This globular cluster is close to **Gamma (γ) Muscae** and has faint stars spread over 18 arc minutes.

NGC 4833: This is a large, faint globular cluster within 1 degree of **Delta (δ) Muscae.** A 4 inch (100 mm) or larger telescope is needed to begin to resolve the cluster into individual stars.

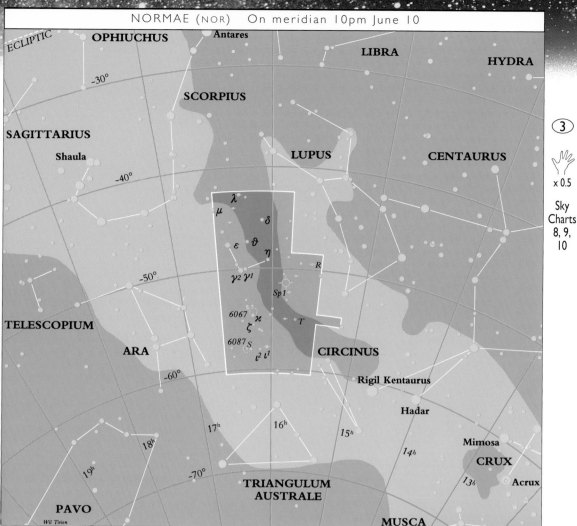

Norma

(NOR-muh) The Square

East of Centaurus and Lupus is a small constellation called Norma, the Square. When he named this group of stars, Nicolas-Louis de Lacaille decided to call it Norma et Regula, the Level and Square, after a carpenter's tools. Since those days, however,

NGC 6067 is a group of 8th magnitude and fainter stars overlying the background of Milky Way stars.

SKYWATCHING TIP

Just because the sky is clear does not necessarily mean that the night is an ideal one for observing. The best nights offer both good transparency and good seeing.

A night without clouds, haze, or light pollution is said to be transparent. However, if the atmosphere above you is turbulent, the images you see through a telescope will wave about like flags in a breeze. When this happens, we say that the seeing is poor. Under such conditions, your telescope will not see the faintest stars, or be able to resolve the double stars that it should. However, a crisp view of planetary detail in good seeing is well worth the wait.

the Regula has been forgotten. The constellation lies alongside Circinus, the Drawing Compass, which he named at the same time.

Set in the southern Milky Way, Norma presents good fields for binoculars, with a number of open clusters. For a small constellation, Norma has also been quite lucky with the appearance of novae: there was one in 1893 and another in 1920.

NGC 6067: This is a small open cluster. Large binoculars or a telescope reveal some 100 stars within a stunning field.

NGC 6087: This is another of Norma's striking open clusters.

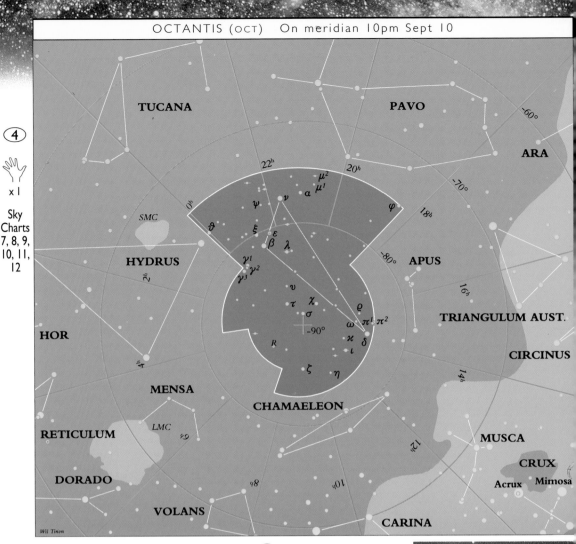

TUCANA

PAVO

ARA

-60°

-70°

22ʰ   20ʰ

μ² μ¹ a

φ 18ʰ

SMC

ψ ν

0ʰ ξ ε

ϑ β λ

HYDRUS

2ʰ

-80°

APUS

γ¹ γ² γ³

ν χ

τ σ ω π¹ π²

-90° ϰ δ ι

R ζ η

16ʰ

TRIANGULUM AUST.

CIRCINUS

HOR

MENSA

CHAMAELEON

14ʰ

MUSCA

RETICULUM

LMC

69

CRUX

DORADO

48

10ʰ

Acrux Mimosa

VOLANS

CARINA

12ʰ

Wil Tirion

Sky Charts 7, 8, 9, 10, 11, 12

4
x 1

Octans

(OCK-tanz)   The Octant

To honor John Hadley's invention of the octant in 1730, Nicolas-Louis de Lacaille formed this south polar constellation and called it Octans Hadleianus. The forerunner of the sextant, the octant was an instrument used for measuring the altitude of a celestial body—an essential device for navigators and astronomers.

The Octant, as drawn by Bode in Uranographia *(1801)*

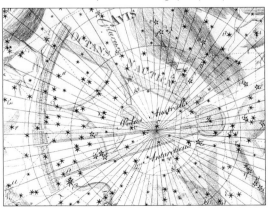

Octans, with the SMC at top left and the stars of Apus to the lower right

👁 **Sigma (σ) Octantis**: This is the south pole star. At magnitude 5.4, it is barely visible to the naked eye on a dark night, so while it does mark the pole, it is not as convenient a marker star as Polaris.

The celestial poles move with time as the axis of the Earth precesses, or wobbles like a top, over some 26,000 years (see p. 87). Sigma (σ) Octantis was at its closest to the pole in about 1870, at just under ½ degree. It will be just over 1 degree from the pole by the year 2000.

In about another 3,000 years, the pole will begin to move through Carina, and it will pass near **Delta (δ) Carinae** in about 7,000 years. At 2nd magnitude, this is the brightest south pole star the Earth ever sees.

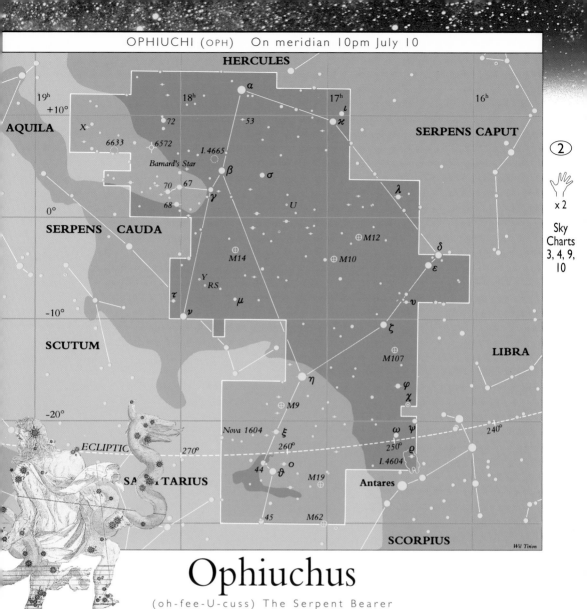

HERCULES

19ʰ
+10°
18ʰ
17ʰ
16ʰ

AQUILA

α

72
53
ι
κ

SERPENS CAPUT

X
6633
6572
I.4665
Barnard's Star
β
σ

λ

2

x 2

70   67
γ
68
U

0°

SERPENS   CAUDA

⊕ M12

M14

⊕ M10

δ
ε

Sky
Charts
3, 4, 9,
10

Y
RS

τ
ν
μ

υ

-10°

SCUTUM

ζ

M107

LIBRA

η

φ
χ

-20°

M9

Nova 1604
ξ
260°

ω   ψ
250°   ϱ

240°

ECLIPTIC
270°

44
ϑ
o

M19

I.4604

SA   TARIUS

Antares

45
M62

SCORPIUS

Wil Tirion

Ophiuchus

(oh-fee-U-cuss) The Serpent Bearer

Ophiuchus, entwined with the constellation of Serpens, covers a large expanse of sky and is filled with points of interest, including some of the Milky Way's richest star clouds. Greek for "serpent bearer", Ophiuchus is usually identified with Asclepius, the god of medicine.

In one legend, Asclepius learned about the healing power of plants from a snake. His medical skills were so great that he could even raise the dead, which was a cause of concern to Hades, god of the underworld. He therefore persuaded Zeus, his brother, to strike Asclepius dead with a thunderbolt. Zeus then placed Asclepius in the sky, in recognition of his healing skills, along with Serpens, his serpent.

On October 9, 1604, Ophiuchus hosted our galaxy's most recent supernova. Known as Kepler's Star, it outshone Jupiter for several weeks.

M 9, 10, 12, 14, 19, and 62: These globular clusters provide a range of examples of different concentrations of stars. M 9 and 14 are rich; M 10 and 12 are looser; M 19 is

oval; M 62 is somewhat irregular in outline. All are visible in binoculars but require a 6 or 8 inch (150 or 200 mm) telescope to do them justice.

RS Ophiuchi: This recurrent nova had outbursts in 1898, 1933, 1958, 1967, and 1985. Its minimum is around magnitude 11.8, rising as high as 4.3 during outbursts.

Barnard's Star: Discovered by E. E. Barnard in 1916, this 9.5 magnitude red dwarf star has the greatest proper motion (apparent motion across the sky) of any known star. Only 6 light years away, it is the nearest star after the Alpha (α) Centauri system.

(Above, left) Ophiuchus is shown holding the Serpent, Serpens, in Bayer's Uranometria (1603). (Right) M 62, one of Ophiuchus's globular clusters, as recorded by an 8 inch (200 mm) telescope

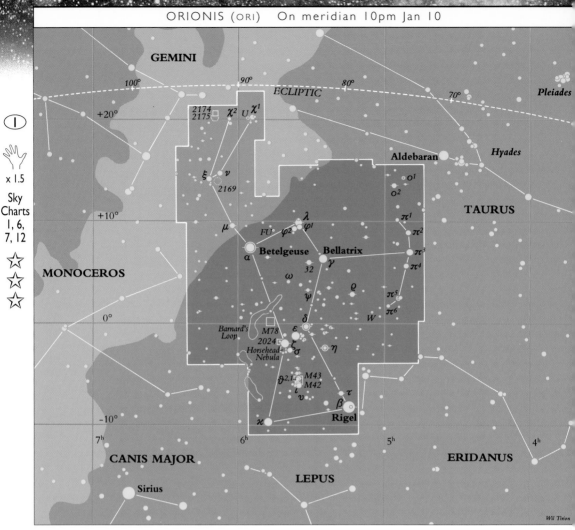

ORIONIS (ORI)   On meridian 10pm Jan 10

GEMINI

ECLIPTIC

Pleiades

TAURUS

Aldebaran

Hyades

MONOCEROS

Betelgeuse   Bellatrix

Barnard's
Loop

M78
2024

Horsehead-
Nebula

M43
M42

Rigel

CANIS MAJOR

ERIDANUS

LEPUS

Sirius

Wil Tirion

Orion

(oh-RYE-un) The Hunter

Orion has been recognized as a distinctive group of stars for thousands of years. The Chaldeans knew it as Tammuz, named after the month in which the familiar belt stars first rose before sunrise. The Syrians called it Al Jabbar, the Giant. To the ancient Egyptians it was Sahu, the soul of Osiris.

In Greek mythology, Orion was a giant and a great hunter. In one story, Artemis, goddess of the Moon and of the hunt, fell in love with him and neglected

Orion, with blue-white Rigel at his foot and red Betelgeuse at his shoulder

her task of lighting the night sky. Her twin brother, Apollo, seeing Orion swimming far out to sea, challenged his sister to hit what was no more than a dot among the waves. Not realizing that this was Orion, Artemis shot an arrow and killed him. Later, when Orion's body was washed up on the shore, she saw what she had done. Inconsolable, she placed his body in the sky, together with his hunting dogs. Her grief explains why the Moon looks so sad and cold.

Orion is a treasure, with Rigel, Betelgeuse, and its three belt stars in a row lighting up the sky from December to April. The stars are arranged so that it is easy to see the figure of a hunter, complete with lion's skin in his right hand and raised club in his left.

Betelgeuse (Alpha [α] Orionis): Betelgeuse (pronounced BET-el-jooze but sometimes corrupted to BEETLE-juice) is fabulous. (Its name comes from the Arabic for "house of the twins", apparently because of the adjacent constellation of Gemini.) A variable star, it varies in magnitude from 0.3 to 1.2 over a

194

The Great Nebula in Orion (M 42) is clearly visible (right) as the central star of Orion's sword, hanging from the three belt stars. The specially processed image (below) reveals details deep in the glowing heart of M 42, with M 43 just above.

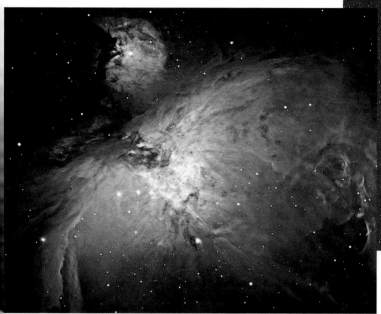

period of almost seven years. However, the semi-regular nature of the variation means that it is often possible to detect changes over just a few weeks.

 Rigel (Beta [β] Orionis): The name Rigel is derived from the Arabic for "foot". This mighty supergiant, which is about 1,400 light years away, is more than 50,000 times as luminous as the Sun.

The Great Nebula (M 42): This star nursery, one of the marvels of the night sky, is also known as the **Orion Nebula**. Plainly visible to the naked eye under a dark sky, it can be clearly seen through binoculars in the city. The swirls of nebulosity spread out from its core of four stars called the **Trapezium,** which power the nebula.

Photographs usually "burn-out" the inner region of the nebula and obscure the Trapezium stars. In 1880, using M 42 as his subject, Henry Draper was the first person to successfully photograph a nebula.

M 43: This is a small patch of nebulosity just north of the main body of the Great Nebula. In fact, the M 42 complex is simply the brightest part of a gas cloud covering the constellation of Orion at a distance of some 1,500 light years.

The Horsehead Nebula (IC 434): Also known as **Barnard 33**, this dark nebula is projected against a background of diffuse nebulosity, alongside the bright belt star **Zeta (ζ) Orionis**. It can be quite difficult to see, usually requiring a dark sky and at least an 8 inch (200 mm) telescope.

NGC 2169: This is a small, bright open cluster of around 30 stars.

The famous Horsehead Nebula, as recorded by the 150 inch (3.9 m) Anglo-Australian Telescope in Australia

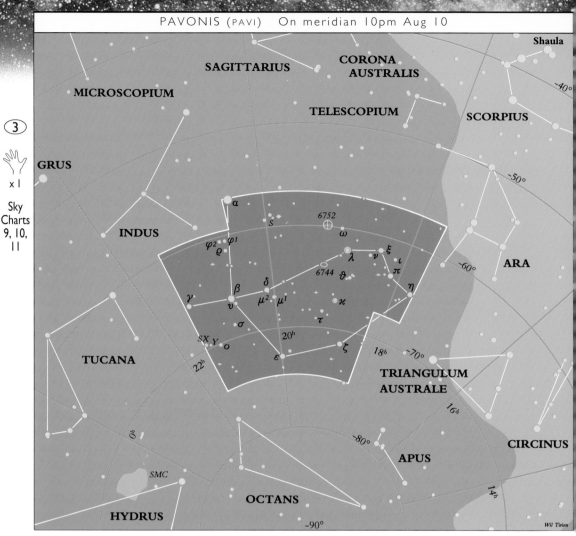

Pavo

(PAH-voh) The Peacock

Pavo, the Peacock, lies not far from the south celestial pole, south of Sagittarius and Corona Australis. It is a modern constellation, devised by Johann Bayer and published in his star atlas of 1603, but he may have been thinking of the mythical peacock that was sacred to Hera in Greek mythology.

Hera suspected that her husband, Zeus, had fallen in love with the mortal Io and had changed his mistress into a white heifer as a disguise. Hera therefore asked Argus Panoptes—a giant with 100 eyes—to guard the heifer. Hermes was sent by Zeus to kill Argus, and when Hera heard of the giant's death she honored him by distributing his eyes over the tail of the peacock.

The Peacock Star (Alpha (α) Pavonis): This star is 150 light years away. It is a binary system whose members orbit each other in less than two weeks, but the pair is too close to separate telescopically.

NGC 6752: A spectacular globular cluster at a relatively close distance of 17,000 light years, this huge family of stars is the third largest globular cluster (in apparent size) after Omega (ω) Centauri and 47 Tucanae.

NGC 6744: This faint but beautiful galaxy is one of the largest known barred spirals. Smaller telescopes reveal only the nuclear regions, a 10 inch (250 mm) telescope being necessary to reveal more.

Large and bright, NGC 6752 is one of the best globular clusters in the sky, but little known because of its southerly location.

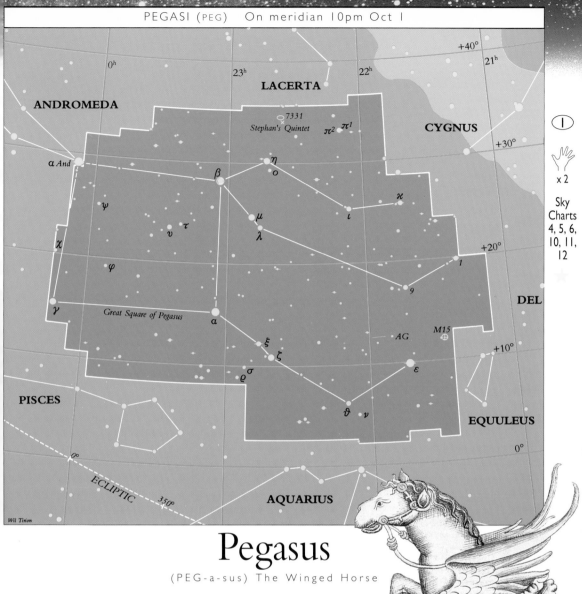

+40°

0ʰ   23ʰ   22ʰ   21ʰ

LACERTA

ANDROMEDA

CYGNUS

+30°

7331
Stephan's Quintet   π² π¹

α And   β   η
                     ο

ψ   ϰ
                    μ   ι
χ   τ   λ
    υ

φ   1   +20°

                         9

γ   Great Square of Pegasus   α   DEL

                         AG   M15   +10°
        ξ ζ
      ρ σ   ε

PISCES   ϑ ν

                    EQUULEUS

                         0°

ECLIPTIC   350°   AQUARIUS

Wil Tirion

I

✋
× 2

Sky
Charts
4, 5, 6,
10, 11,
12

Pegasus

(PEG-a-sus) The Winged Horse

Although this constellation has no really bright stars, it is easy to spot because its three brightest stars, with Alpha (α) Andromedae, form the Great Square of Pegasus.

The winged horse has been found on ancient tablets from the Euphrates Valley, and on Greek coins minted in the fourth century BC. According to Greek legend, when Perseus decapitated the Gorgon Medusa, Pegasus sprang up from her blood. When Pegasus was brought to Mount Helicon, one kick of his hoof caused the spring of Hippocrene to flow—source of inspiration for poets.

Pegasus, as conceived by Domenico Bandini in his fifteenth-century encyclopedia of the universe, Fons Memorabilium Universi

Four of the five galaxies of Stephan's Quintet appear to be interacting, distorting each other, and drawing out long streamers of stars.

M 15: One of the best of the northern sky globular clusters, M 15 is 34,000 light years away. Although it is visible through binoculars as a nebulous patch, in a telescope it is a real showpiece.

NGC 7331: This spiral galaxy is the brightest one in Pegasus, but is still only 9th magnitude.

Stephan's Quintet: This very faint group of galaxies lies ½ degree south of NGC 7331. Even though faint streamers of material seem to connect the largest of the galaxies to the others, detailed study indicates that this galaxy is probably closer to us than they are. These galaxies are not really targets for the beginner, as they need at least a 10 inch (250 mm) telescope to be seen clearly.

197

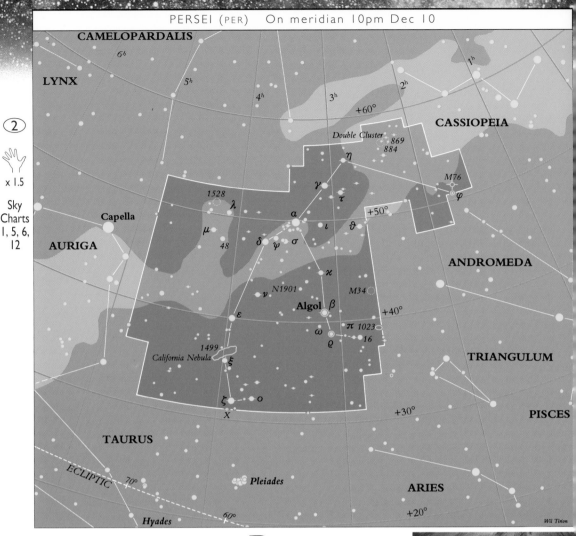

CAMELOPARDALIS

LYNX

CASSIOPEIA

Double Cluster   869   884

M76

1528

Capella

AURIGA

ANDROMEDA

California Nebula 1499

Algol β

M34

N1901

TRIANGULUM

TAURUS

Pleiades

ARIES

PISCES

Hyades

ECLIPTIC

Wil Tirion

2
x 1.5
Sky
Charts
1, 5, 6,
12

Perseus

(PURR-see-us) The Hero

A pretty constellation that straddles the
Milky Way, Perseus is in the northern
skies from July to March. Its stars arc
from Capella, in Auriga, to Cassiopeia.

The son of Zeus and the mortal Danaë,
Perseus's most famous exploit was to kill the
Gorgon Medusa—one of three sisters that were
so terrifyingly ugly that one glance of them
would turn the viewer to stone. Using Athene's
shield as a mirror, he severed Medusa's head, and
the winged horse Pegasus sprang from her blood.

Algol: The star that winks, this is the most
famous of the eclipsing variables. Every 2 days,
20 hours, and 48
minutes, it begins to
drop in brightness
from magnitude 2.1
to 3.4 in an eclipse
lasting 10 hours.

*Perseus, carrying
Medusa's head, with
Pegasus, in a detail
from Rubens' "Perseus
and Andromeda"*

M 34: This
bright open
cluster sits in
the middle of a
rich field of stars.
It is an inter-
esting view
through either binoculars or a telescope.

Double Cluster (NGC 869 and 884) :
Two of the finest examples of open
clusters in the sky, NGC 869 and 884 (h Persei
and Chi [χ] Persei respectively) are magnificent
through binoculars or the low-power field of a
small telescope.

Perseids: One of the best meteor showers,
these meteors, which come from Periodic
Comet Swift-Tuttle, peak on August 11 and 12.

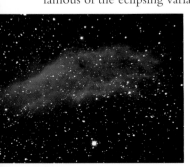

*NGC 1499, the California
Nebula, is not easy to see
as it is large and faint.*

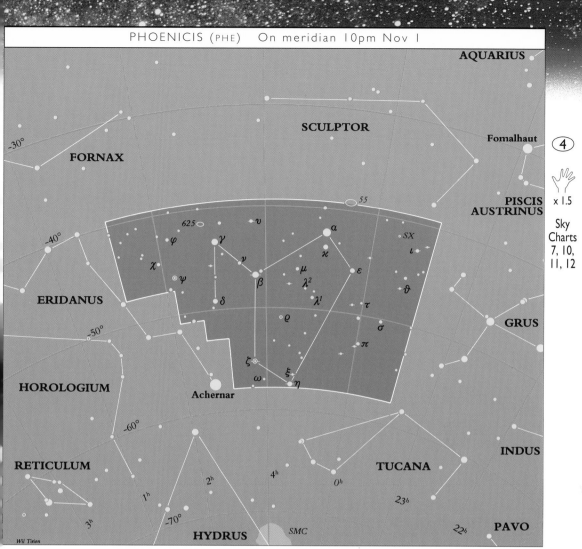

Phoenix

(FEE-nicks) The Phoenix

A fabulous symbol of rebirth, in mythology the Phoenix was a bird of great beauty that lived for 500 years. It would then build a nest of twigs and fragrant leaves which would be lit by the noontime rays of the Sun. The Phoenix would be consumed in the fire, but a small worm would wriggle out from the ashes, bask in the Sun, and quickly evolve into a brand new Phoenix. Depictions of this miraculous bird have been found in ancient Egyptian art and on Roman coins.

Although the constellation first appeared in Bayer's *Uranometria* of 1603, the idea of a Phoenix in the sky goes back to the ancient Chinese who visualized a firebird that was known as Ho-neaou.

SX Phoenicis: The best example of a "dwarf Cepheid" variable, this star changes from magnitude 7.1 to 7.5 and back again in only 79 minutes and 10 seconds! Cepheid periods are very exact. In this case, however, the range varies, with some maxima as bright as 6.7. The variation probably occurs because the star

NGC 625 is a faint (12th magnitude) irregular galaxy, lying 20 to 30 million light years away.

has two different oscillations occurring at once. Such a small range in brightness can be difficult to monitor, requiring very careful comparison with neighboring stars.

199

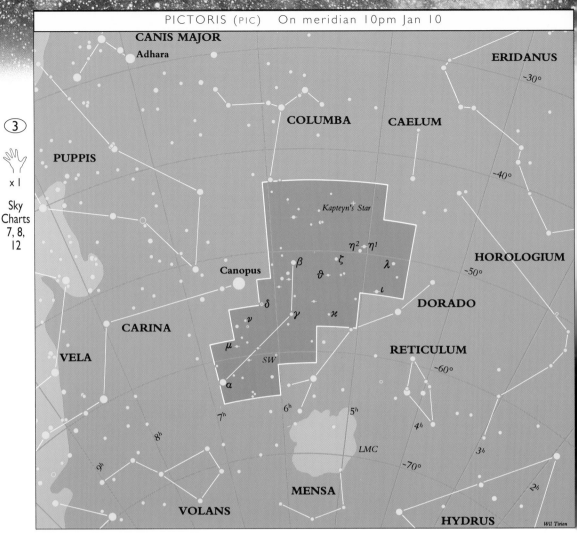

Pictor

(PIK-tor) The Painter's Easel

This southern constellation was originally named Equuleus Pictoris, the Painter's Easel, by Nicolas-Louis de Lacaille. Nowadays its shortened name refers solely to the painter. It is a dull group of stars lying south of Columba and alongside the brilliant star Canopus.

An unusual nova appeared here in 1925. Although it was a bright 2nd magnitude star at the time it was discovered, it continued to brighten until it was almost 1st magnitude. It then began to fade, but brightened again to a second maximum two months later.

Beta (β) Pictoris: This 4th magnitude star is host to a disk of dust and ices which could be a planetary system in formation. The surrounding nebula is only visible using special techniques on large telescopes.

Kapteyn's Star: Only 12.7 light years away, this star was discovered by the famous Dutch astronomer Jacobus Kapteyn in 1897. It moves quickly among the distant background stars, crossing 8.7 arc seconds of sky per year—the width of the Moon every two centuries. At magnitude 8.8, the star is visible through binoculars and small telescopes.

We see the disk of dust and ices around Beta (β) Pictoris nearly edge on. The central circle and dark lines are artifacts of the imaging process.

200

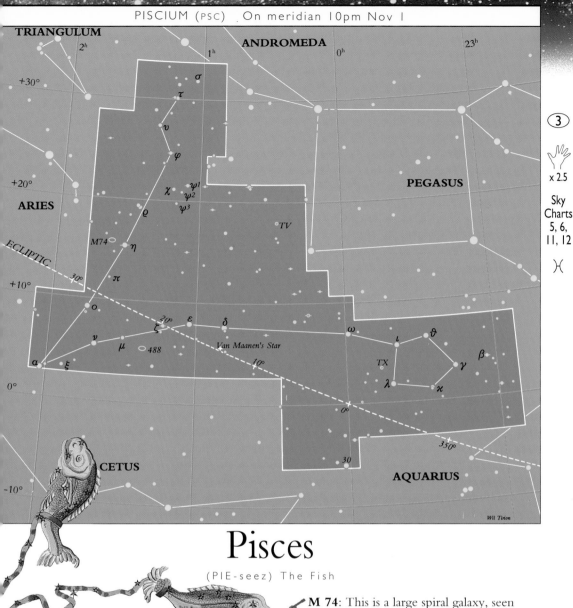

TRIANGULUM
ANDROMEDA
PEGASUS
ARIES
ECLIPTIC
M74
Van Maanen's Star
TV
TX
CETUS
AQUARIUS

Wil Tirion

③

x 2.5

Sky
Charts
5, 6,
11, 12

♓

Pisces

(PIE-seez) The Fish

For thousands of years, this faint zodiacal constellation has been seen either as one fish or two. In Greco-Roman mythology, Aphrodite and her son Heros were at one time being pursued by the monster Typhon. To escape him, they turned themselves into fish and swam away, having tied their tails together to make sure that they would not be parted.

The ring of stars in the western fish, which is beneath Pegasus, is called the Circlet. The eastern fish is beneath Andromeda.

A triple conjunction, in which Jupiter and Saturn appeared close to one another three times in a single year, took place in Pisces starting in August of 7 BC. This rare event is one candidate for a celestial phenomenon which might have been the star of Bethlehem.

Zeta (ζ) Piscium: A beautiful double star of magnitudes 5.6 and 6.5, separated by 24 arc seconds.

M 74: This is a large spiral galaxy, seen face-on, close to **Eta (η) Piscium**. While it is the brightest Pisces galaxy, it is still rather faint and requires a dark sky and an 8 inch (200 mm) telescope or larger to be seen.

Van Maanen's Star: This is a rare example of a white dwarf star that, at magnitude 12.2, can actually be identified in an 8 inch (200 mm) telescope.

(Above, left) The Fish in Urania's Mirror (1825), with their tails tied together in accordance with the legend

(Above) M 74 is faint, but it is a classic example of a spiral galaxy with a prominent nucleus and well-developed spiral arms. 201

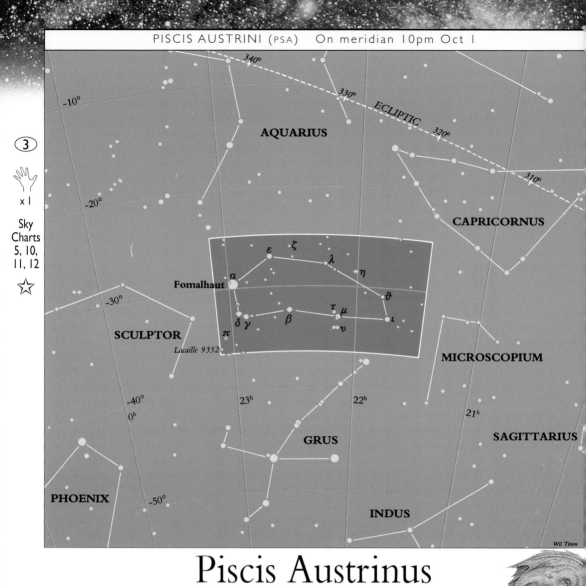

AQUARIUS

ECLIPTIC

CAPRICORNUS

Fomalhaut

SCULPTOR

Lacaille 9352

MICROSCOPIUM

GRUS

SAGITTARIUS

PHOENIX

INDUS

Wil Tirion

Sky Charts 5, 10, 11, 12

Piscis Austrinus

(PIE-sis OSS-trih-nuss) The Southern Fish

Lying to the south of Aquarius and Capricornus, Piscis Austrinus, the Southern Fish, is relatively easy to spot because of its lone, bright star, Fomalhaut, which is often referred to as The Solitary One. For the Persians, 5,000 years ago, this was a Royal Star that had the privilege of being one of the guardians of heaven.

Fomalhaut stands out as a beacon of Piscis Austrinus and the surrounding sky.

Bayer's 1603 depiction of the Southern Fish looks more like a mythical sea monster than any fish you are likely to find in the southern oceans.

Many early charts of the heavens show the Southern Fish drinking water that is being poured from Aquarius's jar.

Fomalhaut: At magnitude 1.2, this star is 22 light years away—close by stellar standards. It is about twice as large as our Sun and has 14 times its luminosity.

Some 2 degrees of arc southward is a magnitude 6.5 dwarf star that seems to be sharing Fomalhaut's motion through space. They are so far apart that it is hard to call them a binary system. Maybe these two stars are all that is left of a cluster that dissipated long ago.

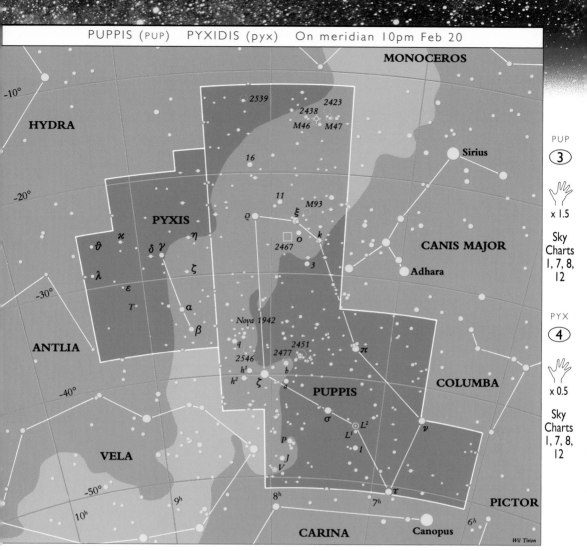

MONOCEROS

-10°

HYDRA

2539
2423
2438
M46   M47

16

Sirius

-20°

11   M93

PYXIS

ϱ
ξ

ϑ κ   δ γ   η

ο
2467
k

CANIS MAJOR

3

Adhara

λ

-30°

ε
τ

α

β

Nova 1942

ANTLIA

q
2546
h¹
h²   ζ

2451
2477   b
a

π

COLUMBA

-40°

PUPPIS

σ

L²
L¹   I

ν

VELA

P

J
V

τ

PICTOR

-50°

9ʰ

8ʰ

7ʰ

6ʰ

CARINA

Canopus

Wil Tirion

PUP
③
×1.5
Sky
Charts
1, 7, 8,
12

PYX
④
×0.5
Sky
Charts
1, 7, 8,
12

Puppis & Pyxis

(PUP-iss) The Stern   (PIK-sis) The Compass

Found just south of Canis Major, Puppis is the stern of the ship *Argo*. It is the northernmost of the constellations that formed the ship. With the Milky Way running along it, Puppis provides a feast of open clusters for binoculars or telescopes. Right alongside is the smaller and fainter constellation of Pyxis, which used to be Malus, *Argo*'s mast, before Lacaille made a ship's compass out of it.

Zeta (ζ) Puppis: This blue supergiant sun is one of our galaxy's largest. About 2,000 light years away, it shines at 2nd magnitude.

L² Puppis: One of the brightest of the red variable stars, L² Puppis varies from magnitude 2.6 to 6.2 over a period of five months.

Bode's illustration of the Compass, from Uranographia (1801)

Look for the faint red ring of the planetary nebula NGC 2438 in the northeast (upper left) quadrant of the cluster M 46.

M 46: A beautiful open star cluster, through small telescopes M 46 is a circular cloud of faint stars the apparent diameter of the Moon. A faint planetary nebula, **NGC 2438**, appears to be a part of the cluster but is not a true member. It is 11th magnitude and 1 arc minute across, needing an 8 inch (200 mm) or larger telescope to be seen well.

T Pyxidis: A recurrent nova with a faint 16th magnitude minimum, T Pyxidis sometimes reaches 7th magnitude during its outbursts, which occur at intervals of 12 to 25 years.

203

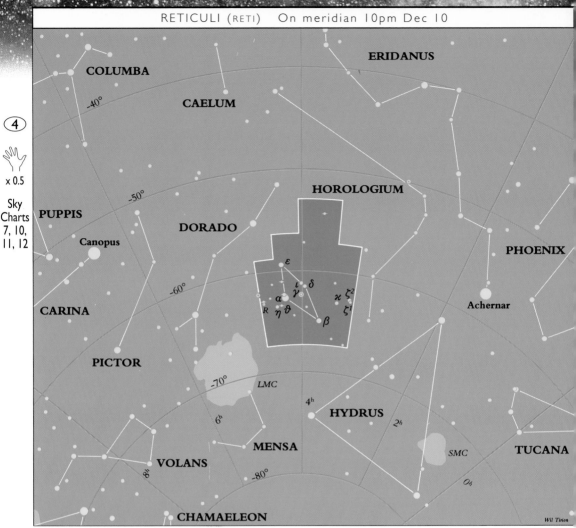

COLUMBA

CAELUM

ERIDANUS

-40°

4

×0.5

Sky
Charts
7, 10,
11, 12

PUPPIS

-50°

DORADO

HOROLOGIUM

PHOENIX

Canopus

ε

-60°

γ ι δ

α κ ζ²

R η ϑ β ζ¹

Achernar

CARINA

PICTOR

-70°   LMC

4ʰ

HYDRUS

2ʰ

6ʰ

MENSA

SMC

TUCANA

VOLANS

8ʰ

-80°

0ʰ

CHAMAELEON

Wil Tirion

Reticulum

(reh-TIK-u-lum) The Reticule

A small constellation of faint stars half way between the bright star Achernar and Canopus, Reticulum was first set up as Rhombus by Isaak Habrecht of Strasburg. De Lacaille changed its name to Reticulum to honor the reticle—the grid of fine lines in a telescope

SKYWATCHING TIP

The joy of skywatching is enriched when shared with young children. As you point out constellations to them, give them time to formulate questions, and don't be afraid to say if you don't know the answers. One eight-year-old girl once told me that the globular cluster M 13 looked like a "little fluffy dog". It wasn't astrophysics, but it signaled the start of her personal relationship with the sky.

eyepiece that aids with centering and focussing. It is occasionally also known as the Net.

R Reticuli: This Mira star is quite red, and at maximum light it shines at about magnitude 7. Over a period of nine months, it drops to magnitude 13, then returns to maximum brightness.

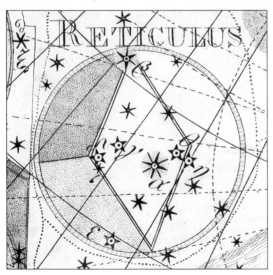

Bode's illustration reflects Reticulum's origin as a rhombus rather than its more recent identification as the scale sometimes found in a telescope eyepiece.

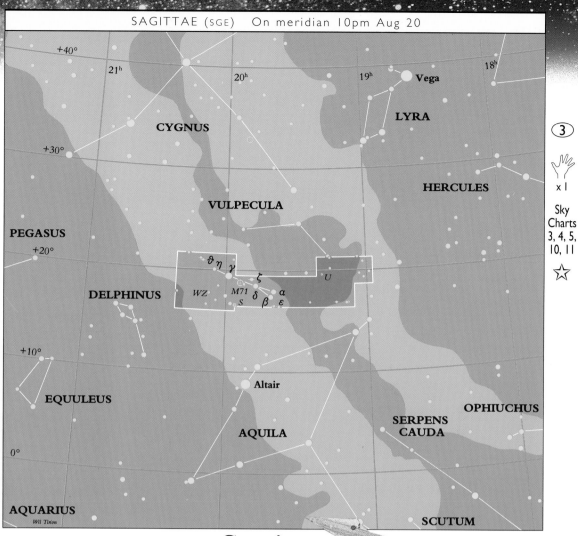

+40°

21ʰ   20ʰ   19ʰ   Vega   18ʰ

CYGNUS   LYRA

+30°

                                         HERCULES   ③

VULPECULA   🖐 x 1

PEGASUS   Sky
+20°   Charts
                ϑ η γ   3, 4, 5,
DELPHINUS   WZ   M71 ⊕ ζ   U   10, 11
                    S δ β ε   ☆
                      α

+10°

                         Altair

0°   SERPENS   OPHIUCHUS
EQUULEUS   CAUDA

              AQUILA

AQUARIUS
Wil Tirion   SCUTUM

Sagitta

(sa-JIT-ah) The Arrow

A lthough only a small constellation, Sagitta is easy to find halfway between Altair in Aquila, and Albireo (Beta [β] Cygni). It really is true to its name—the ancient Hebrews, Persians, Arabs, Greeks, and Romans all saw this group of stars as an arrow. It has been thought of variously as the arrow that Apollo

The Arrow, as depicted by Bayer in his famous star atlas Uranometria (1603)

used to kill the Cyclops; one of the arrows shot by Hercules at the Stymphalian Birds; and as Cupid's dart.

U Sagittae: Every 3.4 days, this eclipsing binary drops from magnitude 6.5 to a minimum of 9.3.

V Sagittae: Although this star is faint, varying erratically from magnitude 8.6 to magnitude 13.9, it is interesting for it alters a little almost every night. It might have been a nova a long time ago.

M 71: A little south of the midpoint of a line joining **Delta (δ)** and **Gamma (γ) Sagittae**, M 71 is a fertile cluster of faint stars. It is now generally regarded as a poor, uncondensed globular cluster, rather than as a rich open cluster.

M 71 is a very loose globular cluster, without the bright central core typical of its kind.

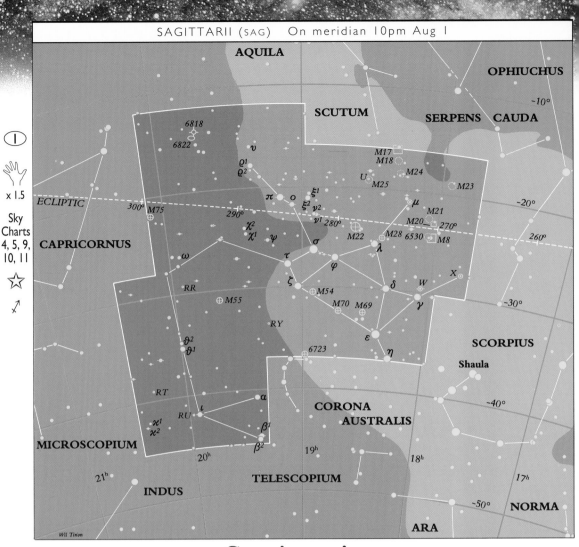

AQUILA

OPHIUCHUS

SCUTUM

SERPENS CAUDA

−10°

6818
6822

M17
M18

υ

M24

M25

M23

U

ECLIPTIC   300° M75

290°

π ο

ξ¹
ξ² ν²

μ

M21

−20°

ν¹ 280°

M20

270°

260°

χ²
χ¹   ψ

σ

λ

M28 6530   M8

CAPRICORNUS

ω

τ

φ

M22

X

−30°

RR

ζ

M54

δ

W

M55

M70 M69

γ

RY

ε

SCORPIUS

6723

η

Shaula

ϑ²
ϑ¹

−40°

RT

α

CORONA
AUSTRALIS

RU ι

β¹

κ¹
κ²

−50°

β²

19ʰ

18ʰ

17ʰ

MICROSCOPIUM

20ʰ

TELESCOPIUM

NORMA

21ʰ

INDUS

ARA

Wil Tirion

Sagittarius

(sadge-ih-TAIR-ee-us) The Archer

This is one of the 12 constellations of the zodiac. The most distinctive aspect of Sagittarius is the group of stars within it that look like a teapot, complete with spout and handle. The charts mark the eastern set of Sagittarius's brighter stars as a quadrilateral figure, which is the teapot's handle. The western group, marked as a triangle, is the spout. The handle also stands by itself as the Milk Dipper.

Not knowing much about teapots, the ancient Arabs thought of the western triangle as a group of ostriches on their way to drink from the Milky Way, and the eastern quadrilateral as ostriches returning from their refreshment.

Sagittarius is generally thought to be a centaur—half man and half horse—and is usually considered to be Chiron, who is also identified with the constellation Centaurus. However, Sagittarius is seen as holding a drawn bow, which is not in character with Chiron, who was known for his wisdom and kindness. Some say that Chiron created the constellation to guide Jason and the Argonauts as they sailed on the *Argo*.

Sagittarius is located on the Milky Way in the direction of the center of the galaxy. Here the band of the Milky Way is at its broadest, although cut by dark bands of dust. It is a treasure trove of galactic and globular clusters, plus bright and dark nebulae.

The Archer, as he appears in a sixteenth-century fresco at the Villa Farnese in Italy

The bright western half of Sagittarius overlaps the star clouds of the Milky Way looking toward the galactic center. The pink glow of the M8 nebula is quite apparent, and is shown close up below. Several other bright features of Sagittarius and Scorpius can also be discerned.

M 22: The Great Sagittarius star cluster is a very large globular—the best of the constellation's many globulars. At magnitude 6.5 it is an easy object to see in binoculars, but a telescope really brings out the cluster's beauty. Only 10,000 light years away, it is one of the closest globulars, and with an 8 inch (200 mm) telescope you should be able to resolve it into seemingly countless stars.

M 23: Just one of many galactic clusters in Sagittarius, M 23 presents over 100 stars in any area the size of the Moon. It is a striking sight in binoculars or in a telescope at low magnification.

The Lagoon Nebula (M 8): This spectacular diffuse nebula envelops the cluster of stars called **NGC 6530**. On a dark night the nebula is visible to the naked eye just to the north of the richest part of the Sagittarius Milky Way. In photographs, the extensive nebula is marked by several tiny dark splotches. Bart Bok identified these as globules in which new stars are being formed.

The Trifid Nebula (M 20): Found only 1 ½ degrees to the northwest of the Lagoon Nebula, the Trifid Nebula is likely to be part of the same complex of nebulosity. It is known as the Trifid because three lanes of dark clouds divide the nebula in the most beautiful way. You should be able to detect these dark lanes with a 6 inch (150 mm) telescope under a good sky.

The Omega Nebula (M 17): Also called the Swan, the Horseshoe, or the Check-mark, this nebula can be seen quite clearly in binoculars and is a stunning sight through a large telescope.

(Above) M 22 is the most striking of the many globular star clusters in Sagittarius.
(Left) The Omega Nebula (M 17) on the border of Sagittarius and Serpens Cauda

207

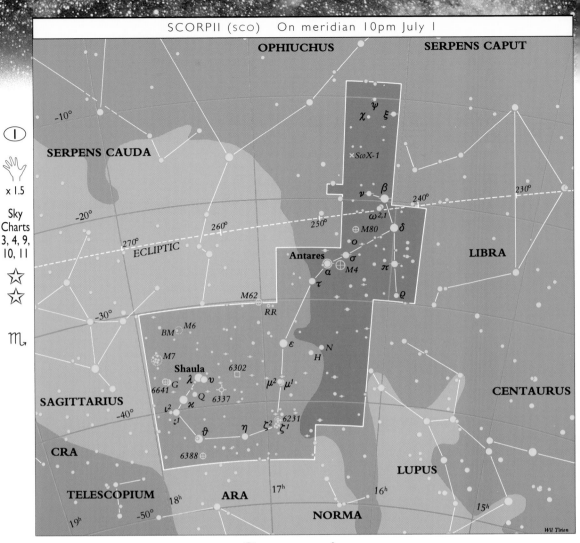

OPHIUCHUS

SERPENS CAPUT

SERPENS CAUDA

-10°

×*ScoX-1*

ψ

χ   ξ

ν   β

240°

230°

-20°

260°

250°

ω²,¹

M80

δ

LIBRA

270°

ECLIPTIC

o

Antares

σ

π

α   M4

τ

ϱ

M62

-30°

RR

BM   M6

ε

N

M7

H

Shaula   6302

μ² μ¹

CENTAURUS

G   λ   ν

6641

SAGITTARIUS

Q   ϰ   6337

-40°

ι²

6231

ι¹   ϑ   η   ζ² ζ¹

CRA

6388

LUPUS

TELESCOPIUM   18ʰ   ARA   17ʰ   16ʰ   15ʰ

-50°

19ʰ   NORMA

Wil Tirion

Scorpius

(SKOR-pee-us) The Scorpion

In Greek mythology, Scorpius is the scorpion that killed Orion. The two constellations are set at opposite sides of the sky, to avoid further trouble between them.

A beautiful constellation of the zodiac, filled with bright stars and rich star fields of the Milky Way, Scorpius really looks like a scorpion, complete with head and stinger. Near the northern end is a line of three bright stars, with red Antares (Greek for "rival of Mars") at its center.

Some 5,000 years ago, the Persians thought of Antares as one of the Royal Stars, a guardian of heaven. The ancient Chinese referred

A thirteenth-century Italian view of Scorpius, with Antares near the center of the Scorpion's body

to Antares's red glow as the "Great Fire" at the heart of the Dragon of the East. Another Chinese legend refers to Antares and its two attendants as the Ming T'ang—the "Hall of Light" or the "Emperor's Council Hall".

Antares: The Romans called this mighty star Cor Scorpionis, meaning "heart of the scorpion", a title the French also use—Le Coeur de Scorpion. Thought to be about 520 light years away, Antares is a red supergiant some 600 million miles (1,000 million km) across and is 9,000 times more luminous than the Sun. However, with a mass only 10 or 15 times that of the Sun, it is not very dense. Its insides might be like a very hot vacuum.

Beta (β) Scorpii: This is a double star whose 2.6 and 4.9 magnitude components are 13.7 arc seconds apart, making resolution possible in a 2 inch (50 mm) telescope.

M 4: "There are several M 4s," said Walter Scott Houston, a highly insightful observer who died in 1993 at the age of 81. What he meant was that this strange globular cluster has a

This three-color composite image shows Antares enmeshed in nebulosity at lower left, with the globular cluster, M 4, to its west (right). The Rho (ρ) Ophiuchi nebulosity (IC 4604) appears at the top.

different appearance with each instrument you use. Binoculars show a fuzzy patch of light; a small telescope shows a large patch of mottled haze; and 4 or 6 inch (100 to 150 mm) instruments begin to show the individual stars. This is one of the best globulars for viewing in small telescopes.

The Butterfly Cluster (M 6): The stars of this large, bright open cluster really resemble a butterfly when they are viewed at high power.

M 7: This large, bright open cluster, lying southeast of M 6, needs to be seen through the large field of view of binoculars to be fully appreciated.

NGC 6231: Half a degree north of **Zeta (ζ) Scorpii**, this bright open cluster lies in a rich region of the Milky Way. It is best surveyed in binoculars or at very low power in a telescope.

M 80: This small, bright globular cluster can be seen in binoculars but needs a 10 inch (250 mm) telescope to resolve its stars.

Scorpius X-1: This is a close binary star in which one star expels gas onto a dense neighbor which could be a white dwarf, a neutron star, or a black hole. It is a bright X-ray source, but appears visually as a faint 13th magnitude star.

(Above) The relatively loose globular cluster M 4, is easily found just over 1 degree west of Antares.

(Left) Scorpius, with splendid Antares, is one of the few constellations that looks like the creature after which it is named. Its tail overlies rich Milky Way star clouds, which are dimmed and cut by dark dust lanes.

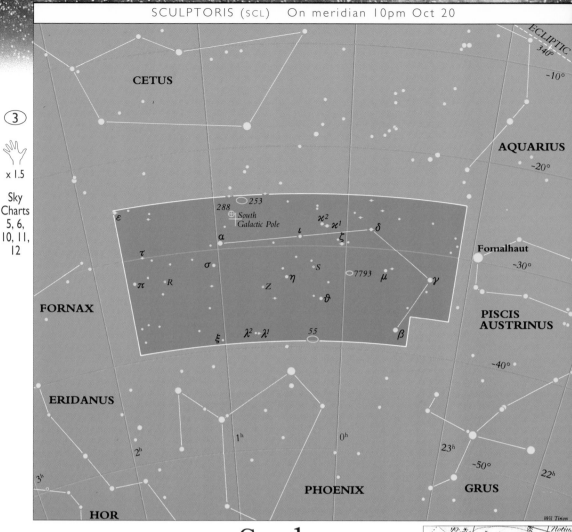

Sculptor

(SKULP-tor) The Sculptor

This constellation, originally named L'Atelier du Sculpteur (the Sculptor's Workshop) by Nicolas-Louis de Lacaille, lies to the south of Aquarius and Cetus. Since that time, its name has been shortened to Sculptor. It contains little of interest, its most significant feature being a small cluster of nearby spiral galaxies.

Johann Bode's depiction of Sculptor in Uranographia *(1801) seems more like the workshop than the sculptor himself.*

🔭 **NGC 253**: For a small telescope user, this galaxy is one of the most satisfying, especially for observers in the Southern Hemisphere. It is very large and is viewed almost edge-on. It was discovered by Caroline Herschel one night in 1783, while she was searching for comets. It appears as a thick streak in binoculars and begins to show the texture evident in photographs when larger instruments are used.

🔭 **NGC 55**: This is another very fine edge-on galaxy, similar to NGC 253. It is distinctly brighter at one end than the other when seen through an 8 inch (200 mm) telescope.

Both NGC 55 and 253 are members of the Sculptor Group, an arrangement of several galaxies that might be our Local Group's nearest neighbor in the cosmos.

NGC 253, at magnitude 7, is a large, bright spiral galaxy. Ten million light years distant, it is the largest member of the Sculptor Group of galaxies.

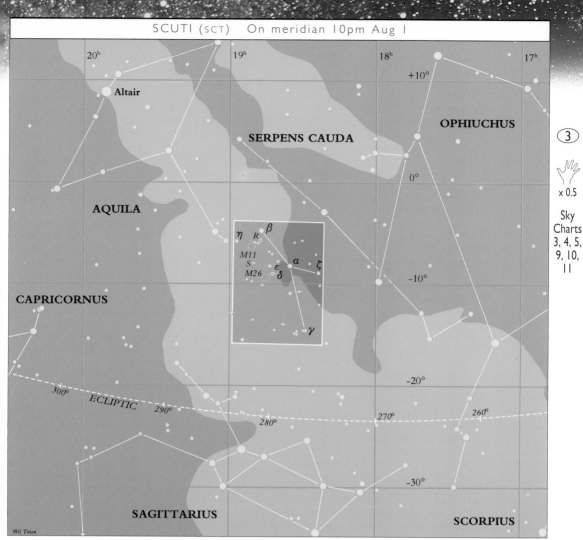

20ʰ   19ʰ   18ʰ   17ʰ

+10°

Altair

SERPENS CAUDA

OPHIUCHUS

③

0°

AQUILA

η   R   β
M11
S
M26   ε δ   α   ζ
-10°

CAPRICORNUS

γ

-20°

300°   ECLIPTIC   290°   280°   270°   260°

-30°

SAGITTARIUS   SCORPIUS

Wil Tirion

x 0.5

Sky
Charts
3, 4, 5,
9, 10,
11

Scutum

(SKU-tum)   The Shield

Although Scutum is not a large constellation and has no bright stars, it is not difficult to find in a dark sky because it is the home of one of the Milky Way's most dramatic clouds of stars. Johannes Hevelius created the constellation at the end of the seventeenth century, giving it the name Scutum Sobiescianum (Sobieski's Shield) in honor of King John Sobieski of Poland, after he had successfully fought off a Turkish invasion in 1683.

R Scuti: This is a semiregular RV Tauri-type variable star. It changes from magnitude 5.7 to 8.4 and back over a period of about five months.

The Wild Duck Cluster, M 11, is quite a concentrated clutch of stars and presents a glorious view in almost any telescope.

The Wild Duck Cluster (M 11): This spectacular open cluster is clearly visible in binoculars, rewarding in a small telescope, and stunning in an 8 inch (200 mm) one. One of the most compact of all the open clusters, the presence of a bright star in the foreground adds to its beauty.

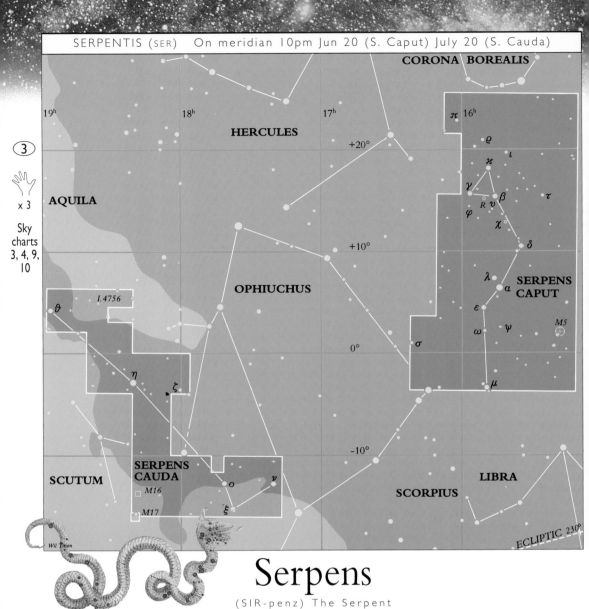

CORONA BOREALIS

HERCULES

+20°

AQUILA

(3)

x 3

Sky charts 3, 4, 9, 10

π 16ʰ

ρ ι

ϰ

γ β τ
R υ
φ
χ

δ

λ
α SERPENS
CAPUT

ε

ω ψ M5

+10°

OPHIUCHUS

19ʰ   18ʰ   17ʰ

I.4756

ϑ

η

ζ

0°

σ

μ

-10°

SCUTUM

SERPENS
CAUDA
□ M16

□ M17

ο ν

ξ

SCORPIUS

LIBRA

ECLIPTIC 230°

Wil Tirion

Serpens

(SIR-penz) The Serpent

This is the only constellation that is divided into two parts. The head (Serpens Caput) and the tail (Serpens Cauda) are separated by the constellation of Ophiuchus, the Serpent Bearer. At one time both the Serpent and the Serpent Bearer formed a single constellation. Serpens was familiar in ancient times to the Hebrews, Arabs, Greeks, and Romans.

R Serpentis: A Mira star almost midway between **Beta (β)** and **Gamma (γ) Serpentis**, this variable has a bright maximum of 6.9. It fades to about 13.4, although it sometimes can become fainter. Its period is about one year.

The Eagle Nebula derives its name from the evocatively shaped dark dust features visible at the center of the photograph.

(Above) Another of the coiling snakes in the sky, Serpens' full length is partially concealed by the figure of Ophiuchus. (Left) M 5, a 5th magnitude globular cluster in Serpens Caput is one of the finest in the sky.

M 5: This very striking globular cluster is about 26,000 light years away.

The Eagle Nebula (M 16): Through an 8 inch (200 mm) or larger telescope on a dark night, this combination of nebula and star cluster is quite stunning. But you can still enjoy the sight of the cluster in smaller telescopes.

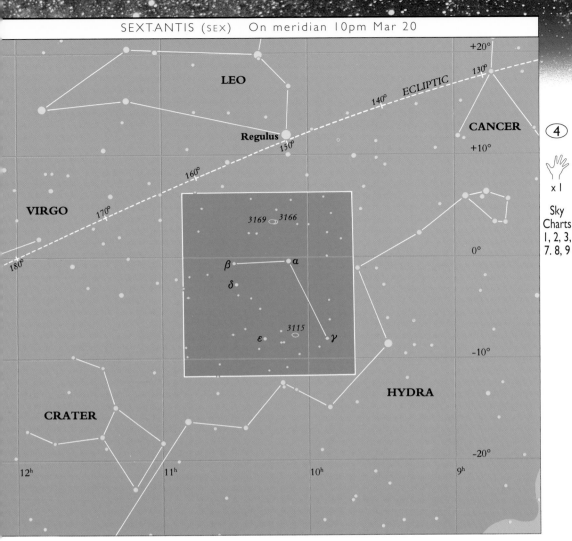

+20°

LEO

130°

ECLIPTIC

140°

CANCER

④

Regulus

150°

+10°

160°

×1

Sky
Charts
1, 2, 3,
7. 8, 9

VIRGO

170°

3169 ⊙ 3166

0°

180°

β•   •α

δ•

3115

-10°

ε•   ⊙   •γ

HYDRA

CRATER

-20°

12ʰ   11ʰ   10ʰ   9ʰ

Sextans

(SEX-tanz) The Sextant

Sextans Uraniae, now known simply as Sextans, was the creation of Johannes Hevelius. He chose this name for the constellation to commemorate the loss of the sextant he once used to measure the positions of the stars.

Along with all his other astronomical instruments, the sextant was destroyed in a fire that took place in September 1679. "Vulcan overcame Urania," Hevelius remarked sadly, commenting on the fire god having defeated astronomy's muse.

Placed between Leo and Hydra, Sextans' brightest star is barely visible to the unaided eye at magnitude 4.5. Despite this, in ancient times the Chinese chose one of the faintest stars in Sextans to represent Tien Seang, the Minister of State in Heaven.

The galaxy NGC 3115 is a rarity, in that the way it appears to the eye in amateur telescopes is similar to the view that is provided in photographs, although perhaps the pointed ends cannot be seen so well.

The Spindle Galaxy (NGC 3115): Because we see this 10th magnitude galaxy almost edge-on, it appears to be shaped like a lens. Unlike many faint galaxies, the Spindle Galaxy gives quite satisfying views at high power. In classification, it seems to be somewhere between an elliptical and a spiral.

213

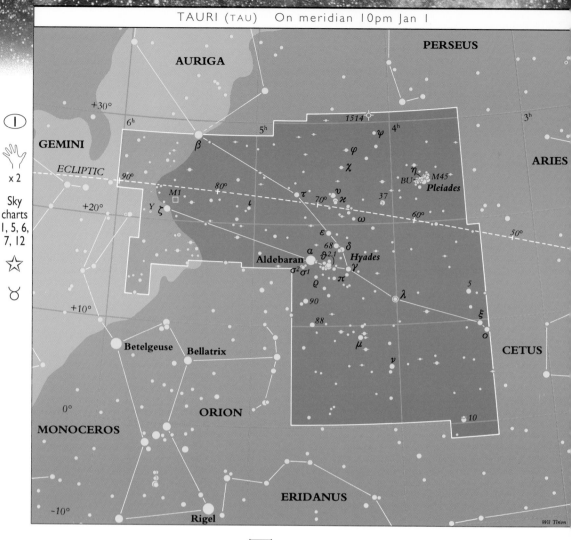

PERSEUS

AURIGA

+30°

6ʰ

GEMINI

ECLIPTIC   90°

+20°

M1

γ ζ

+10°

Betelgeuse   Bellatrix

0°

MONOCEROS

-10°

Rigel

1514

ψ   4ʰ

φ

χ

η   M45

BU   Pleiades

τ   70°   ν   37

ϰ

ω   60°

ε

68   δ

α   ϑ2,1   Hyades

Aldebaran   σ2 σ1   γ

ρ   π

90

88

μ

ν

ORION

ERIDANUS

ARIES

3ʰ

50°

5

ξ

ο

CETUS

10

Wil Tirion

Taurus

(TORR-us) The Bull

Just northwest of Orion, Taurus is a prominent northern constellation containing two of the largest star clusters we can see—the Hyades and the Pleiades. From the time of the Chaldeans, some 5,000 years ago, this constellation has been seen as a bull.

Bulls have been worshipped since ancient times as symbols of strength and fertility, and they appear in numerous myths. The ancient Egyptians worshipped Apis, the Bull of Memphis, a real bull that was thought to be an incarnation of Osiris. The Israelites worshipped the Golden Calf. Carvings of winged bulls stood at the gates of Assyrian palaces.

In classical times, the Greeks saw the constellation as Zeus disguised as a bull. The story goes that

Zeus fell in love with the beautiful Europa, daughter of Agenor, king of Phoenicia. One day, playing at the water's edge, Europa's attention was caught by a majestic white bull—Zeus in animal form—grazing peacefully among her father's herd. The bull knelt before her as she approached it, so she climbed on its back, wreathing flowers around its horns. Springing to its feet, the bull then took off into the sea and swam to Crete, where Zeus made Europa his mistress. One of their three sons, Minos, later became king of Crete. Only the forequarters of the bull are visible in the constellation, as it is emerging from the waves.

A depiction of the zodiacal figure of Taurus from a fresco in the palace of Schifanoia in Ferraro, Italy, by Francesco del Cassa (1436–78)

214

Many a night I saw the

Pleiades, rising thro' the

mellow shade,

Glitter like a swarm of fire-flies

tangled in a silver braid.

Locksley Hall, ALFRED LORD TENNYSON, (1809–92), English poet

They remained in the sky, staying close together. The seventh sister is hard to see because she really wants to go back to Earth, and her tears dim her luster.

On a reasonably dark night, you should be able to see at least six of the stars in the Pleiades with the naked eye; under good conditions, you might be able to see as many as nine. Containing more than 500 stars in all, the Pleiades is about 410 light years away and covers an area four times the size of the full Moon. It is best seen with binoculars.

The Seven Sisters of the Pleiades make a spectacular photograph, bathed in a blue reflection nebula. (Left) A chart of the Pleiades region

👁 **The Hyades**: Like the Pleiades, this is also an open cluster, but it is so close to us (only 150 light years away) that even when viewed with the naked eye the stars appear to be spread out. The stars of the Hyades form the bull's head.

👁 **Aldebaran (Alpha (α) Tauri)**: This is an orange giant and is the brightest star in Taurus. Its name means "the follower" (of the Pleiades) in Arabic. Only 60 light years away, it marks the eye of the bull.

🔭 **The Crab Nebula (M 1)**: This nebula marks the site of the supernova seen in 1054. It is clearly visible through a 4 inch (100 mm) telescope on a dark night as an oval glow, but this only hints at the complex structure that can be seen in high-magnification photographs.

👁 **The Pleiades (M 45)**: Also known as the Seven Sisters, this is the most famous open star cluster in the sky and forms the bull's shoulder. Greek legend tells that the sisters called for help from Zeus when they were being pursued by Orion. Zeus turned them into doves and placed them in the sky. Alcyone (Eta [η] Tauri) is the most dazzling sister. She is accompanied by Maia (20 Tauri); Asterope I and II (the double star 21 Tauri); Taygeta (19 Tauri); Celaeno (16 Tauri); and Electra (17 Tauri). Finally, there is Merope (23 Tauri), a star surrounded by a beautiful cloud of cosmic grains producing a blue reflection nebula. Atlas (or Pater Atlas, 27 Tauri) and Pleione (Mater Pleione, 28 Tauri) represent the girls' father and mother.

In a Native American tale, the Pleiades are even youngsters who, on a walk through the sky, lost their way and never made it home.

M 1, the famous Crab Nebula, is an oval glow 5 arc minutes across in small telescopes.

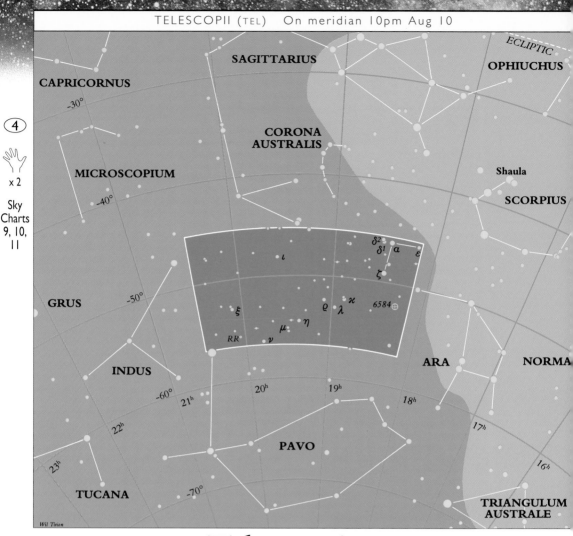

CAPRICORNUS

SAGITTARIUS

OPHIUCHUS

ECLIPTIC

-30°

CORONA
AUSTRALIS

Shaula

MICROSCOPIUM

SCORPIUS

-40°

δ²
δ¹ α
ε

ζ

GRUS   -50°

ξ   ϱ ϰ   6584
λ

μ η
RR   ν

ARA   NORMA

INDUS

-60°   21ʰ   20ʰ   19ʰ   18ʰ

22ʰ   17ʰ

PAVO

23ʰ   16ʰ

TUCANA   -70°

TRIANGULUM
AUSTRALE

Wil Tirion

4

x 2

Sky
Charts
9, 10,
11

Telescopium

(tel-eh-SKO-pee-um) The Telescope

Originally bearing the name Tubus
Telescopium, this constellation was
created by Nicolas-Louis de Lacaille
during the eighteenth century to honor the
invention of the telescope. It was the only large
telescope in space until the launching of
the Hubble Space Telescope in 1990.
Telescopium is surrounded by Sagittarius,
Ophiuchus, Corona Australis, and Scorpius, and
de Lacaille "borrowed" a number of stars from
these constellations in order to create it.

RR Telescopii: Although this star is
normally too faint for small telescopes, it is
one of the most interesting novae
on record. Before 1944, this star
varied over about 13 months
between 12.5 and 15th magnitude,
but in that year it began a rise to magnitude 6.5 that took some five years. As
the nova declined in the following years,

*NGC 6584 is a 9th magnitude globular cluster about 6 arc
minutes across. This image was recorded using a 10 inch
(250 mm) telescope.*

it still displayed its original 13-month period.
It is thought that the star may be a binary
system, in which a large red star is responsible
for the minor variations that take place, and
a smaller, hotter star puts on the nova part of
the performance.

*A simple refracting telescope honors the importance of
the astronomical instrument in Urania's Mirror (1825).*

216

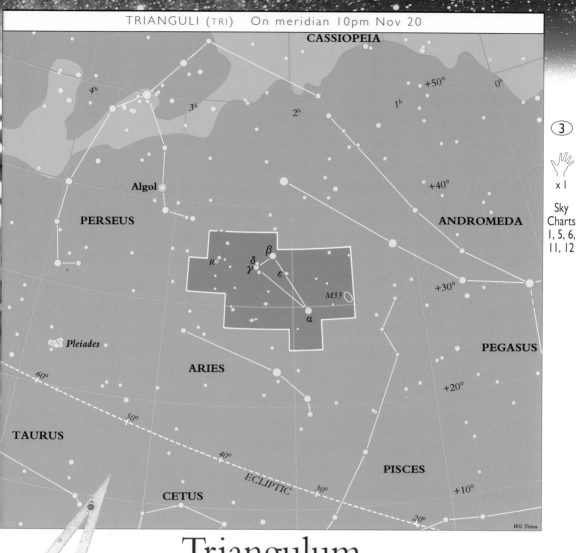

CASSIOPEIA

+50°   0ʰ

4ʰ   3ʰ   2ʰ   1ʰ

+40°

Algol

PERSEUS   ANDROMEDA

β
R°   δ
γ   ε
M33   +30°

α

Pleiades   +20°

ARIES   PEGASUS

60°

50°

TAURUS   PISCES

40°   +10°

ECLIPTIC   30°
CETUS   20°

Wil Tirion

③

🖐
x 1

Sky
Charts
1, 5, 6,
11, 12

Triangulum

(tri-ANG-gyu-lum) The Triangle

Triangulum is a small, faint constellation extending just south of Andromeda, near Beta (β) and Gamma (γ) Andromedae. Despite its lack of distinction, the group of stars was known to the ancients, and because of its similarity to the Greek letter delta (Δ) it was sometimes called Delta or Deltotum. It has been associated with the delta of the River Nile and has also been connected with the island of Sicily, which is shaped like a triangle. The ancient Hebrews gave it the name of a triangular musical instrument.

👁 **The Pinwheel Galaxy (M 33)**: This galaxy is one of the brightest and biggest members of our Local Group and we have a front row view because it appears face-on. The galaxy is listed at magnitude 5.5 but its light is spread out over such a large area that it is notoriously difficult to see. Although it can be seen by the naked eye on very clear nights, you need a dark sky and binoculars to see a fuzzy glow larger than the apparent diameter of the

Moon. A telescope with a wide field of view will also show the galaxy, but one with a narrow field will show nothing at all.

(Above, left) The Triangle was represented in a simple form by Bayer in Uranometria (1603).
(Below) M 33 is a loosely wound spiral and the next largest galaxy in the Local Group of galaxies after the Milky Way and Andromeda (M 31) galaxies.

217

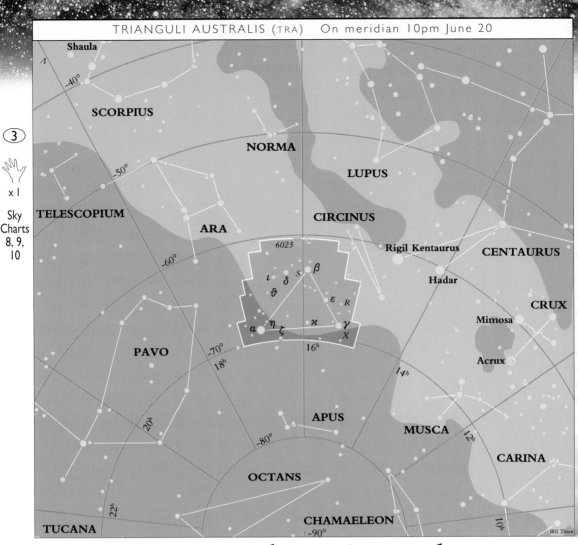

3

x 1

Sky
Charts
8, 9,
10

Triangulum Australe

(tri-ANG-gyu-lum os-TRAH-lee) The Southern Triangle

A simple three-sided figure deep in the southern sky, Triangulum Australe first appeared in Johann Bayer's great atlas *Uranometria* in 1603. It lies just south of Norma, the Level, and east of Circinus, the Drawing Compass—tools used by woodworkers and navigators on early expeditions to the Southern Hemisphere.

R Trianguli Australis: One of several Cepheids in the constellation, this interesting variable alters by about a magnitude—from 6.0 to 6.8. Because it is a Cepheid variable, we know its period precisely, which is 3.389 days. For Cepheids with this rapid variation, magnitude estimates at least once a night are worthwhile.

S Trianguli Australis: Another bright Cepheid variable, S varies from magnitude 6.1 to 6.7 and back over a period of 6.323 days.

NGC 6025: This is a small open cluster of about 30 stars of 9th magnitude, with fainter background stars.

A small, but easily identified constellation, Triangulum Australe is the southern counterpart of Triangulum. Alpha (α) Centauri, in the lower right of the picture, far outshines the constellation's three brightest stars.

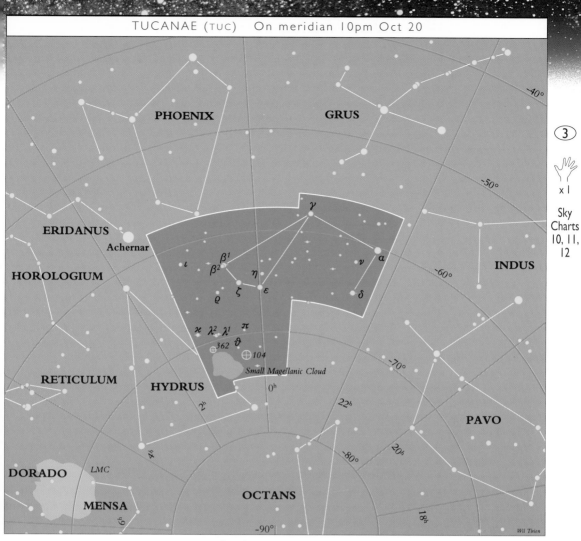

Tucana

(too-KAN-ah) The Toucan

Johann Bayer first published this constellation as Toucan, which in time became Tucana, the Latin form. Toucans are larger members of the genus *Ramphastos*—brightly colored, large-billed birds related to woodpeckers that are found in tropical America. From the earliest drawings, Tucana sat on the Small Magellanic Cloud, one of the two closest galaxies to the Milky Way, tending it like an egg.

👁 **47 Tucanae (NGC 104)**: From its perch 16,000 light years away, this glorious globular cluster shines brightly at magnitude 4.5. Although it is a

The Small Magellanic Cloud (SMC), with the spectacular globular cluster 47 Tucanae alongside. NGC 362, a much smaller and fainter globular, is just north (above) of the SMC.

naked eye object under dark conditions, a 4 inch (100 mm) or larger telescope really brings out the best in this cluster, which competes with Omega (ω) Centauri for the title of the most splendid globular cluster in the entire sky. It is more centrally condensed than its rival in Centaurus.

👁 **The Small Magellanic Cloud (SMC)**: A member of our Local Group, this galaxy is visible to the naked eye on a good night, with 47 Tucanae alongside. A little less than 200,000 light years away, the cloud is some 30,000 light years wide.

The Toucan, as seen in Johann Bode's Uranographia *(1801)*

219

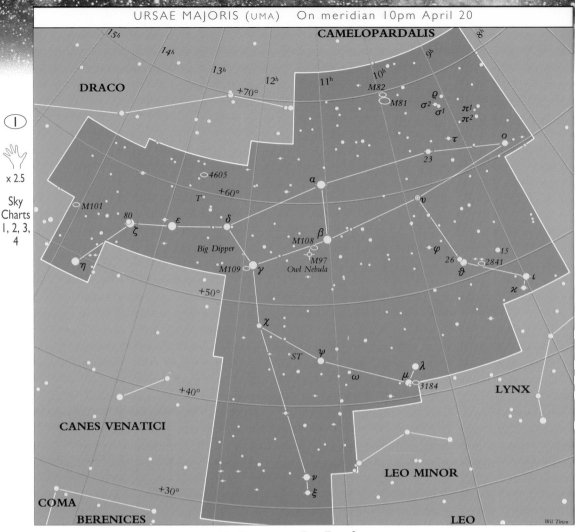

CAMELOPARDALIS

DRACO

M81

M82

M101

Big Dipper

M108

M97
Owl Nebula

M109

CANES VENATICI

LYNX

COMA
BERENICES

LEO MINOR

LEO

Wil Tirion

Ursa Major

(ER-suh MAY-jer) The Great Bear

One of the oldest of the constellations, Ursa Major, The Great Bear, is also perhaps the best known, and numerous legends have been associated with it. Particularly famous is the group of seven stars that make up what is commonly known as the Big Dipper, or the Plough.

In a Cherokee legend, the handle of the Big Dipper represents a team of hunters pursuing the bear from the time he is high in the sky in spring

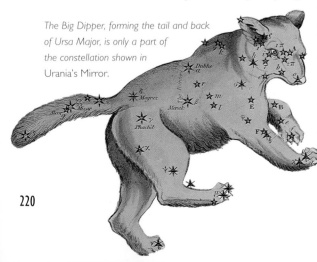

The Big Dipper, forming the tail and back of Ursa Major, is only a part of the constellation shown in Urania's Mirror.

until he sets on fall evenings. At the start of each evening, the bear, with his hunters, has moved a little farther west in the sky.

The Iroquois of Canada's St Lawrence River Valley and the Micmacs of Nova Scotia have a more elaborate story. Represented by the Dipper bowl, the bear is hunted by seven warriors. Each spring the hunt begins when the bear leaves Corona Borealis, his den. The bear isn't killed until fall, and the skeleton remains in the sky until the following spring. Then a new bear emerges from Corona Borealis and the hunt begins again. Instead of a bear, the Sioux of central North America see a long-tailed skunk.

According to one Chinese legend, the stars of the Big Dipper form a bushel measure to deliver food in fair amounts to the population in times of famine. The ancient Hebrews also saw a bushel measure.

The early Britons saw the Big Dipper forming King Arthur's chariot, whereas in Germany people pictured the group of stars as a wagon and three horses. The Romans viewed

220

(Left) M 101, a loosely wound spiral galaxy, is one of the largest and brightest galaxies in the sky.
(Below) Ursa Major is the third largest constellation in the sky, but the Big Dipper is clearly the most prominent part.

it as a team of seven oxen, harnessed to the pole and driven by Arcturus.

In the Greek legend, Zeus and Callisto, a mortal, had a son called Arcas. Hera, Zeus's jealous wife, turned Callisto into a bear, and one day, while out hunting, her son, not knowing that the bear was his mother, almost killed her. Zeus rescued Callisto, placing both her and her son, whom he also turned into a bear, in the sky together. Callisto is Ursa Major and Arcas is Ursa Minor.

Mizar (Zeta [ζ] Ursae Majoris) and Alcor: This is the famous apparent double star in the middle of the Dipper's handle. It is separated by 12 arc minutes and is thus possible to see as a pair by naked eye. Mizar is itself a true binary star, separated by 14 arc seconds.

M 81: This spiral galaxy can be easily seen through binoculars, even when observing in the city, and it is dramatic when observed under good conditions. The oval disk becomes more apparent with increasing telescope size.

M 82: This is a long, thin peculiar galaxy, just ½ degree from M 81. It appears as a thin, gray nebulosity in a 4 inch (100 mm) telescope, but begins to show some detail in an 8 inch (200 mm) or larger one. Even in large telescopes or photographs, however, it is not clear what type of galaxy this is.

M 101: This large, spread-out spiral galaxy is visible through small telescopes if the sky is dark enough. It needs a wide field and a low-power eyepiece. At 16 million light years, it is one of the closer spiral galaxies to the Milky Way.

The Owl Nebula (M 97): This is an oval planetary nebula that takes the shape of an owl when it is seen in a 12 inch (300 mm) telescope. It is large and therefore dim, and a 3 inch (75 mm) or larger telescope is needed to find it.

M 81—a spectacular galaxy— is probably a fair representation of how the Milky Way galaxy would look from the outside.

221

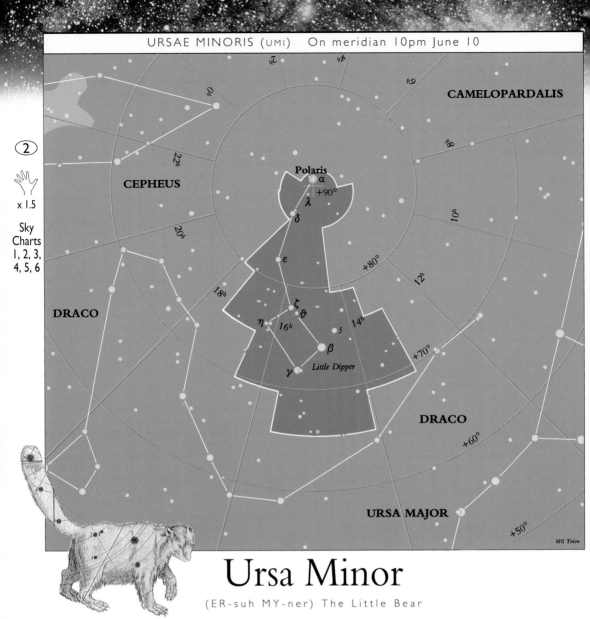

CAMELOPARDALIS

CEPHEUS

Polaris
α
+90°
λ
δ

+80°

ε

18ʰ

η　16ʰ　ζ
5　14ʰ
β
γ　Little Dipper

+70°

DRACO

DRACO

+60°

URSA MAJOR

+50°

Wil Tirion

②

✋
x 1.5

Sky
Charts
1, 2, 3,
4, 5, 6

Ursa Minor

(ER-suh MY-ner)　The Little Bear

Also known as the Little Dipper, Ursa Minor looks somewhat like a spoon whose handle has been bent back by a playful child. This group of stars was recognized as a constellation in 600 BC by the Greek astronomer Thales.

The Little Bear, according to Greek legend, is Arcas, son of Callisto—Ursa Major, the Great Bear. Placed in the heavens by Zeus, he and his mother follow each other endlessly around the north celestial pole.

👁 **Polaris (Alpha [α] Ursae Minoris)**: The pole star for the Northern Hemisphere, this Cepheid variable is currently almost 1 degree from the exact pole. Precession of the Earth's axis will carry the pole to within about 27 arc minutes of Polaris around the year 2100, and then it will start to move away again.

Polaris is 820 light years away, with a 9th magnitude companion some 18½ arc seconds away. Splitting this pair is an interesting test for a 3 inch (75 mm) telescope.

While touring England's Lake District in 1819, a gravely ill John Keats was thinking of Polaris when he wrote these lines:

> Bright Star, would I were steadfast as thou art—
> Not in lone splendor hung aloft the night
> And watching, with eternal lids apart,
> Like Nature's patient, sleepless Eremite.

(Above, left) The Little Bear, as seen by Bayer in Uranometria *(1603), the first "modern" star atlas*

(Below) The relatively faint stars of Ursa Minor swing around Polaris as the sky rotates during the night.

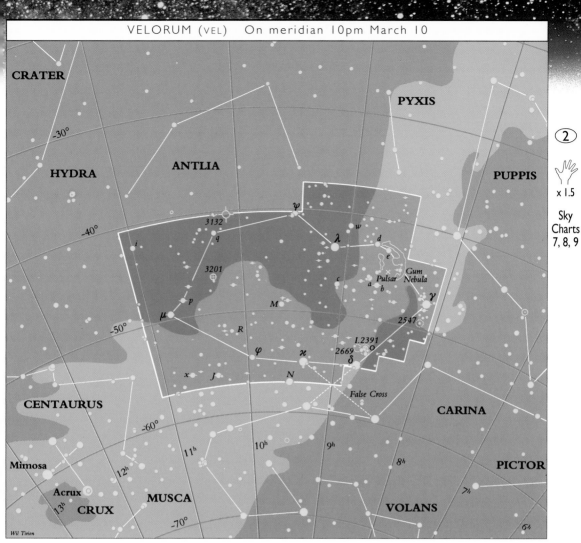

CRATER

PYXIS

ANTLIA

HYDRA

PUPPIS

②
x 1.5
Sky
Charts
7, 8, 9

-30°

ψ

3132

-40°

i

q

λ

w

d

e

3201

c

a
b

Pulsar   Gum
Nebula

p

M

γ

μ

R

2547

-50°

φ

ϰ

2669   I.2391
δ   O

CENTAURUS

x   *J*

N

False Cross

CARINA

-60°

Mimosa   11ʰ

10ʰ   9ʰ

8ʰ

PICTOR

12ʰ

7ʰ

Acrux

MUSCA

VOLANS

CRUX   13ʰ   -70°

6ʰ

Wil Tirion

Vela

(VEE-lah) The Sail (of Argo)

This constellation, along with Carina (the Keel), Puppis (the Stern), and Pyxis (the Compass), once formed part of a huge group of stars in the southern skies known as Argo Navis, the Ship Argo. This was the vessel that Jason and the Argonauts sailed in on their search for the Golden Fleece. Argo Navis was divided up by Nicolas-Louis de Lacaille in the 1750s, sharing its stars between the four resulting constellations. This left Vela with no stars designated alpha (α) or beta (β).

👁 **The False Cross**: **Delta (δ)** and **Kappa [κ] Velorum**, together with **Epsilon (ε)** and **Iota (ι) Carinae**, make up a larger but fainter version of the Southern Cross which is known as the False Cross.

Gamma (γ) Velorum: This double star is resolvable in a steady pair of binoculars. The primary is a Wolf-Rayet star—very hot and luminous.

🔭 **NGC 3132**: This bright planetary nebula accompanies the many clusters in Vela, but lies right on the border with Antlia. Being 8th

A satellite trail cuts across a small portion of the extensive nebulosity forming the Vela Supernova remnant.

magnitude and almost 1 arc minute across, it is considered the southern version of Lyra's Ring Nebula, but with a much brighter central star.

223

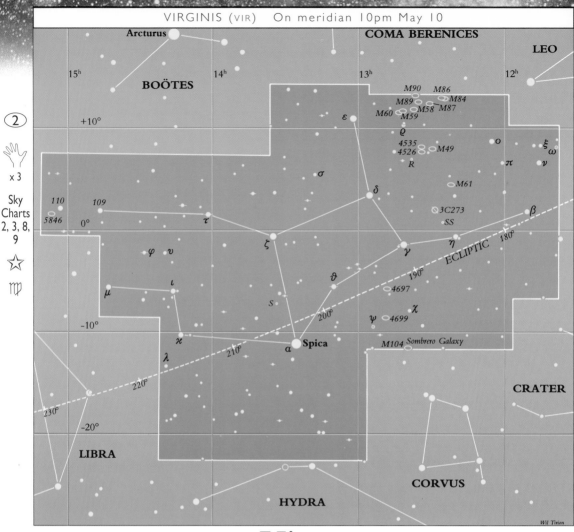

Arcturus

COMA BERENICES

LEO

BOÖTES

15ʰ   14ʰ   13ʰ   12ʰ

M90   M86
M89   M84
M60 M58 M87
M59

+10°

ε

ϱ
4535
4526   M49
R

o

ξ
ω
π   ν

σ

M61

δ

3C273
SS

β

110   109
5846   0°

τ

ζ

γ

η   ECLIPTIC   180°

φ   υ

ϑ

190°

ι

4697

200°

-10°

μ

S

χ

ψ   4699

ϰ

210°

α   Spica

M104 Sombrero Galaxy

λ

220°

CRATER

230°

-20°

LIBRA

CORVUS

HYDRA

Wil Tirion

Virgo

(VER-go) The Maiden, The Virgin

Virgo is the only female figure among the constellations of the zodiac and has been thought to represent a great array of deities since the beginning of recorded history. Among others, she has been identified with the Babylonian fertility goddess Ishtar; Astraea, the Roman goddess of justice; and Demeter, the Greek goddess of the harvest (Roman, Ceres). Virgo is usually shown either holding an ear of wheat or carrying the scales of Libra, the adjoining constellation.

The lovely Virgo, as she is featured in Urania's Mirror of 1825

Spica (Alpha [α] Virginis): A bright white star, Spica is the ear of wheat Virgo is holding. The star is almost exactly 1st magnitude, although it has a slight variation. It is 220 light years away and has more than 2,000 times the luminosity of the Sun.

Porrima (Gamma [γ] Virginis): This is one of the best double stars in the sky, each component shining at magnitude 3.7. At 3 arc seconds separation, the pair is still easy to separate, but by the year 2017 they will appear much closer together.

THE REALM OF THE GALAXIES

Scattered throughout Virgo and Coma Berenices are more than 13,000 galaxies. Known as the Virgo Cluster or Coma-Virgo Cluster, this mighty club of distant systems of stars repays sweeping with a small, wide-field telescope on a dark night. To see much detail usually requires an 8 inch (200 mm) or larger telescope. Listed here are some of the galaxies, but you are likely to see many more.

2

x 3

Sky Charts 2, 3, 8, 9

☆

♍

M 49: This elliptical galaxy is one of the brightest galaxies in the Virgo cluster. It is slightly bigger and brighter than M 87.

M 84 and 86: These two elliptical galaxies are close enough to be seen in the same low-power telescope field. On a dark night, an 8 inch (200 mm) telescope will show several smaller galaxies in the same view.

M 87: This elliptical galaxy is one of the mightiest galaxies we know. Through a small telescope it appears as a bright patch of fuzzy light about a magnitude brighter than M 84 and 86. Interestingly, larger telescopes don't show a great deal more. In the professional size range, however, more details do emerge. With a 60 inch (1.5 m) telescope, for example, you can see a jet emerging from the galaxy's center, and photographs with a 200 inch (5 m) telescope have shown more than 4,000 globular clusters on the galaxy's outskirts.

The Sombrero Galaxy (M 104): Although this galaxy is quite a distance south of the main concentration of galaxies, it seems to be gravitationally attracted to the swarm and so is thought to be a part of it. The brightest of the Virgo galaxies, a dark lane cuts along its equator, making it look a little like a Sombrero hat in an 8 inch (200 mm) telescope.

The Sombrero Galaxy presents a bright, 8th magnitude glow, just 8 arc minutes across, but readily seen in smaller telescopes.

Some of the brighter members of the stunning Virgo Cluster, dominated in this view by M 86 (center) and M 84 (right), both of them elliptical galaxies. The majority of galaxies in the Cluster are about 65 million light years from Earth.

3C273 Virginis: This is the brightest known quasar, but being only 13th magnitude, an 8 inch (200 mm) telescope is needed to identify it. Lying 3 billion light years away, it is the most distant object amateurs are likely to see in their telescopes.

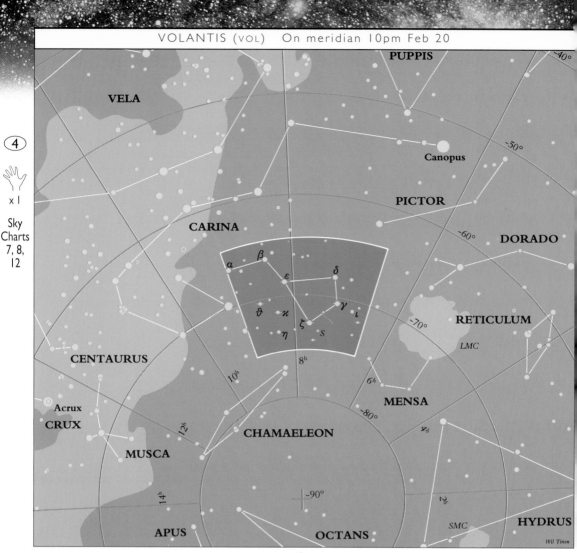

PUPPIS

VELA

Canopus

-40°

-50°

PICTOR

DORADO

-60°

CARINA

α   β
ε   δ
ϑ   κ   γ   ι
ζ   S
η

RETICULUM

-70°

LMC

CENTAURUS

8ʰ

10ʰ

6ʰ

MENSA

-80°

Acrux

CRUX

12ʰ

CHAMAELEON

MUSCA

4ʰ

14ʰ

-90°

2ʰ

SMC

HYDRUS

APUS

OCTANS

Wil Tirion

4

x 1

Sky
Charts
7, 8,
12

Volans

(VOH-lanz) The Flying Fish

The constellation of Piscis Volans, the Flying Fish, lies south of Canopus, and was introduced by Johann Bayer in his *Uranometria* of 1603. It is now known only as Volans. Sailors in the south seas had reported seeing schools of flying fish, which may have been the inspiration for the name. The pectoral fins of these fish are as large as the wings of birds and they glide across the water for distances of up to ¼ mile (400 m).

SKYWATCHING TIP

For long exposure photography of stars and planets, it is important to align the telescope mounting more precisely than for visual observing. This will save you major guiding headaches later. There are several alignment methods, and some companies even manufacture special alignment telescopes that can be attached to your mount's polar axis.

NGC 2442 is an 11th magnitude barred spiral galaxy, seen nearly face-on. Its faintness and 6 arc minute size make it a target for a 12 inch (300 mm) telescope.

S Volantis: A Mira star, S Volantis usually has a maximum magnitude of 8.6, but it has occasionally risen to 7.7. Its faint minimum averages 13.6. The star completes its cycle in a little less than 14 months.

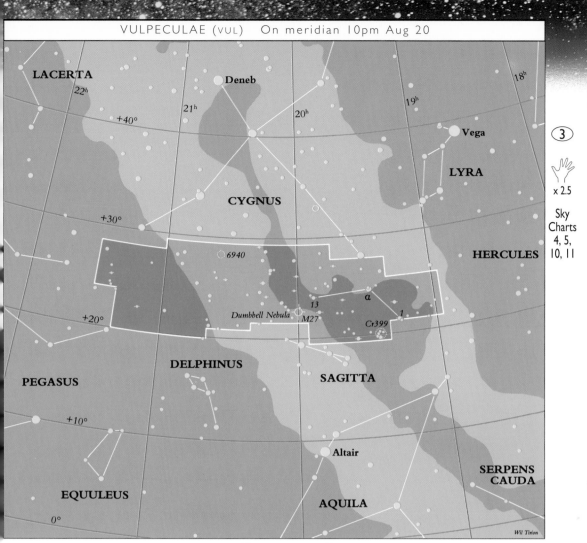

LACERTA

22ʰ

Deneb

21ʰ

+40°

20ʰ

19ʰ

18ʰ

Vega

CYGNUS

LYRA

+30°

6940

HERCULES

③

☌

x 2.5

Sky
Charts
4, 5,
10, 11

13

α

+20°

Dumbbell Nebula   M27

1

Cr399

DELPHINUS

SAGITTA

PEGASUS

+10°

Altair

SERPENS
CAUDA

EQUULEUS

AQUILA

0°

Wil Tirion

Vulpecula

(vul-PECK-you-lah) The Fox

This constellation, invented by Johannes Hevelius in 1690, is without an exciting story or a moral tale. Hevelius's name for it was Vulpecula cum Anser, the Fox with the Goose, but now the constellation is simply referred to as the Fox.

The Dumbbell Nebula (M 27): This is one of the finest planetary nebulae in the sky and is well suited to small telescopes. Bright and large, it is easy to find just north of **Gamma (γ) Sagittae**. Being 7th magnitude, it can be found through binoculars, but it appears only as a faint nebulous spot. If you use a small telescope, you can make out its odd shape. A larger telescope will reveal its 13th magnitude central star.

Urania's Mirror (1825). shows Vulpecula, the Fox, with the goose which he has since lost!

The Dumbbell Nebula is a sight for any telescope. The larger the instrument that is used, the greater the details revealed in the nebula's 5 arc minute disk.

Although the gases in this nebula are expanding at the rate of some 17 miles (27 km) per second, there will be no noticeable change in the nebula's appearance within a human lifetime.

227

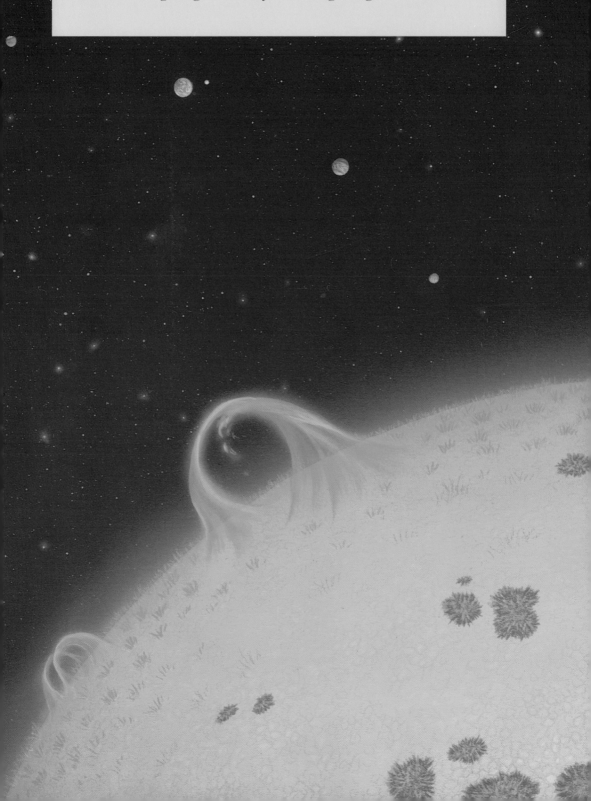

CHAPTER SIX
A TOUR *of the* SOLAR SYSTEM

Spacecraft have revealed the planets to be individual worlds.
A telescope can bring us closer to our nearest neighbors,
giving us some fascinating insights.

THE BIRTH *of* *the* SOLAR SYSTEM

According to one theory, an ancient supernova began the process that resulted in the creation of our Solar System about 4.6 billion years ago.

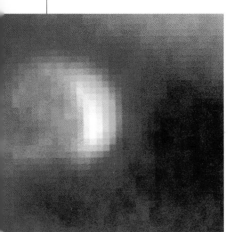

NEW SOLAR SYSTEM? *(left) A very young star (red) in the Orion Nebula, still surrounded by dust, as seen by the Hubble Space Telescope*

HOW THE PLANETS FORMED *(right) Triggered perhaps by an exploding star, the planets formed from a disk of gas and dust surrounding the Sun.*

I magine a large, cold, dark cloud that has rested in space for a great length of time. Somewhere nearby, a star became unstable as it ran out of fuel, then blew itself apart, catapulting much of its contents into the cloud. As a result, the cloud was enriched with some heavier elements

He, who through vast

immensity can pierce,

See worlds on worlds

comprise one universe,

Observe how system into

system runs, What other

planets circle other suns, . . .

May tell why Heaven has

made us as we are.

An Essay on Man,
ALEXANDER POPE (1688–1744),
English poet

from the supernova, including carbon—the basis of life—and began to collapse. As it shrank, it began to rotate faster and its particles clumped together. Most of the material gravitated toward the center, where a mighty protosun began to grow and heat up.

Was this how the Sun began to form? For some reason, a gas cloud did begin to collapse to form the Sun, but what happened to the rest of the cloud as the Sun grew stronger? Some scientists believe that grains of material from the disk consolidated into solid lumps of material. Because there were so many of these bodies, they kept colliding, growing into larger bodies called protoplanets, which became the planets we see today, while some of the leftover small pieces evolved into comets.

According to other theorists, there were many proto-planets that increased in size for a time, but they broke up as a result of repeated collisions with smaller objects and each other. Only the few largest ones were left to consolidate and cool, becoming the nine planets we know today.

The disk of material, known as the accretion disk, continued to rotate, and temperatures soared at its center. As its core—the Sun—"turned on", it blew away the remnants of the cloud, leaving an infant Solar System consisting of a group of small, warm inner planets nestled close to the Sun; some larger, cold outer planets; and small, frozen comets at the outer edges—plus a good deal of rocky debris. And all this took place quite quickly, cosmically speaking. It is estimated that it may have been no more than 100 million years between the time the cloud began to collapse and the moment the Sun ignited.

RELATIVE DISTANCES OF PLANETS FROM SUN *(below) From left to right: Sun, Mercury, Venus, Earth, Mars, Asteroid Belt, Jupiter, Saturn, Uranus, Neptune, and Pluto. The illustration on pp. 228–29 features a close-up view of the planets and how they compare in size with one another.*

THE TURBULENT SUN

The Sun rules our Solar System. It is responsible for our being here—for growing our food and for keeping us warm.

The Sun is a sphere of gas, its heat generated by a nuclear furnace at its center where hydrogen is fused to helium. Nuclear fusion releases huge amounts of energy, which eventually escape the Sun through its turbulent photosphere—the surface that we see. Temperatures at the photosphere hover around 10,500 degrees Fahrenheit (5,800 degrees Celsius), far cooler than the 27 million degrees Fahrenheit (15 million degrees Celsius) at the core.

APOLLO *(above) the Greek god who came to be associated with the Sun*

Above its surface, the Sun has a complex atmosphere which consists of the chromosphere and corona. Although the corona extends into interplanetary space, it is almost as hot as the core.

SUNSPOTS

Although the ancient Chinese recorded dark areas on the Sun 2,000 years ago, it was Galileo who first understood that these sunspots are formed on the Sun, carried across its surface as it rotates, and later dissipated.

In 1828, more than two centuries later, Henrich Schwabe, in Dessau, Germany, began searching for a possible planet named Vulcan as it supposedly passed between the Earth and the Sun, but instead

SUN WARNING *Never look at the Sun through binoculars or an unfiltered telescope (see p. 65).*

discovered that the numbers of sunspots rose and fell in a cycle.

Over an average of 11 years, activity on the Sun, as seen in the sunspots, rises and falls, the most recent peak being in 1989. On Earth, echoes of this sunspot cycle have been found in the width of growth rings in trees.

FACT FILE

Distance from Earth:
93 million miles
(150 million km)

Sidereal revolution period:
365.26 days

Mass (Earth = 1): 333,000

Radius at equator (Earth = 1):
109

Apparent size:
32 arc minutes

Sidereal Rotation period (at equator): 25.4 days

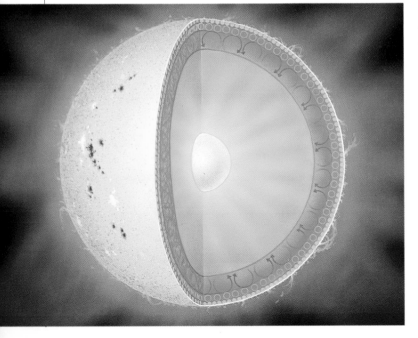

THE SUN'S STRUCTURE *The Sun's visible surface, with its dark spots and looping prominences (pink), overlies boiling convection layers. Deeper down, energy streams out from the nuclear powerhouse at the core (white). Above the surface lies the thin chromosphere (pink), merging into the corona.*

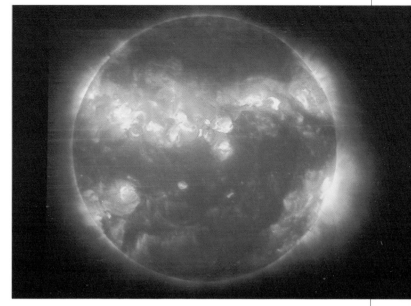

Give me the splendid silent sun with all his beams full-dazzling!

WALT WHITMAN (1819–92)
American poet

Typically, a sunspot consists of a dark region (the umbra) surrounded by a lighter region (the penumbra), although it is quite common to find more than one umbra with a single penumbra. Spots arise where the Sun's magnetic field is concentrated, impeding the flow of energy. Actually a sunspot is not all that dark. The umbra is about 3,600 degrees Fahrenheit (2,000 degrees Celsius) cooler than the photosphere and only looks dark relative to its brilliant surroundings.

OBSERVING THE SUN
Sunspots come in many shapes and sizes, often in groups. A really large group may be 60,000 miles (100,000 km) across—eight times the diameter of the Earth! Even much smaller sunspots are easy to see with a small telescope, provided you do it safely.

As the Sun rotates, the groups of sunspots march across its surface. Over a period of about 10 days, a group will travel from one edge to the other. Watching, drawing, or photographing sunspots as they change is a fascinating area of skywatching to complement your night-time observations. Ironically,

SUNSPOTS *(right) seen a day apart near sunspot maximum in 1989. Note the motion across the disk and changes in detail in just one day.*

THE X-RAY SUN *The bright X-ray emission from the corona revealed in this image, taken from a small rocket, is found to overlie active sunspot regions on the surface.*

the problem with daytime observing is caused by the Sun itself. It warms the ground and the air around you, causing air turbulence and therefore worse seeing than is generally encountered during the night.

FOG CORONA *(left) The true corona of the Sun is faint and overwhelmed by the brilliance of the disk of the Sun, but local fog has produced this brightly colored "corona".*

233

THE MOON:
EARTH'S INTIMATE PARTNER

As inspiring for a poet as it is challenging to a scientist,

the Moon is our closest neighbor in space.

DIANA *The Roman goddess identified with the Moon is usually portrayed as a huntress, as in this sixteenth-century painting by Orazio Gentileschi.*

For an idea of the sort of battering the Earth has received in the past 4 billion years, take a look at the pitted and scarred surface of the Moon. All you need is a pair of binoculars or the smallest of telescopes.

Since the Earth has always shared the same neighborhood as the Moon, but has a much greater surface area, we can be certain that our planet has been hit a good deal more frequently. Over time, the records of these numerous impacts have been eroded on Earth, but they stand out clearly on the Moon as it has no atmosphere.

HOW WAS THE MOON FORMED?

The Earth is unique among the inner planets in having a large natural satellite—the Moon—and there are several theories as to how this came about. Currently popular is the idea that about 4.5 billion years ago some large object struck the Earth, blasting out material that eventually collected together in orbit about the Earth. The pieces struck each other repeatedly

MAN ON THE MOON *Astronaut James Irwin, lunar module, and lunar rover of the Apollo 15 mission, on the desolate surface of the moon*

THE FULL MOON *seen from Earth is dominated by dark "oceans" and "seas" and the bright ray crater Tycho (near the bottom) in the Moon's highlands.*

Marduk bade the moon come forth; entrusted night to her, made her creature of the dark, to measure time; and every month, unfailingly, adorned her with a crown.

Translated from *Enuma Elish*, seventh-century BC tablets upon which Assyro-Babylonian legends concerning the creation of the world are recorded

and melted as they coalesced, eventually cooling off to become the Moon. Over time, the Moon has steadily increased its distance from the Earth, as a result of the same forces that cause tides in the oceans on Earth.

For 500 million years after its formation, the Moon and the rest of the Solar System were bombarded by the pieces of debris that still remained. The final phase of heavy cratering resulted in the formation of large impact basins that subsequently became flooded with lava, forming areas of smooth, dark rock. These dark regions are known as maria (singular, mare) or "seas". Mare Imbrium ("Sea of Showers"), for example, is a basin 600 miles (1,000 km) wide, ringed by mountains, that was formed about 3.9 billion years ago and sometime later inundated with lava. The smooth lava surfaces of all the seas bear only light cratering—the result of more recent impacts.

The remainder of the Moon's surface, unaffected by lava flooding, forms the Moon's bright highlands. These areas are intensely cratered, with their pulverised surface considerably brighter than the maria. Craters range from tiny pits to large craters with central mountain peaks and walled plains, such as Clavius, which is more than 140 miles (225 km) across.

MOUNTAINS AND CRATERS
The lunar highlands, as seen in imagination (left), and from orbit by Apollo spacecraft (center), and the lava plain of Mare Imbrium (below).

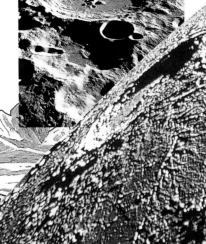

FEATURES OF THE MOON

1	Albategnius	10	Aristoteles	19	Endymion	28	Jura Mts
2	Aliacenis	11	Arzachel	20	Eratosthenes	29	Kepler
3	Alphonsus	12	Atlas	21	Eudoxus	30	Langrenus
4	Alps	13	Autolychus	22	Fracastorius	31	Longomontanus
5	Anaxagoras	14	Catharina	23	Gassendi	32	Maginus
6	Apennine Mts	15	Caucasus Mts	24	Grimaldi	33	Mare Cognitum
7	Archimedes	16	Clavius	25	Hercules	34	Mare Crisium
8	Aristarchus	17	Cleomedes	26	Hipparchus	35	Mare Foecunditatis
9	Aristillus	18	Copernicus	27	Janssen	36	Mare Frigoris

FEATURES OF THE MOON

37	Mare Humorum	46	Oceanus Procellarum	55	Schikard
38	Mare Imbrium	47	Petavius	56	Schiller
39	Mare Nectaris	48	Piccolomini	57	Sinus Iridum
40	Mare Nubium	49	Pitatus	58	Stöfler
41	Mare Serenitatis	50	Plato	59	Straight Wall
42	Mare Smythii	51	Posidonius	60	Theophilus
43	Mare Tranquillitatis	52	Ptolemaeus	61	Tycho
44	Mare Vaporum	53	Purbach	62	Walter
45	Maurolychus	54	Pythagoras	63	Werner

Apollo landing Sites:
- 11 Apollo 11
- 12 Apollo 12
- 14 Apollo 14
- 15 Apollo 15
- 16 Apollo 16
- 17 Apollo 17

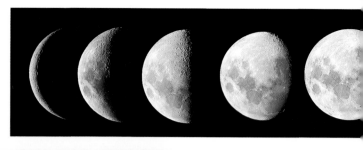

PHASES OF THE MOON, *from crescent to gibbous, to full and back, over one month. Here the Moon is oriented as seen from the Southern Hemisphere.*

EARTHSHINE *faintly illuminates the darkened disk of the Moon, while the sunlit crescent is overexposed.*

The Moon's rotation is locked to its period of revolution around the Earth, and so we never see its far side, but photographs from space have revealed a similar picture there, although there appear to be no large maria.

CRATERS AND RAY SYSTEMS

If you use a pair of binoculars or a small telescope to look at the Moon when it is near full, you will see the bright crater known as Tycho, which is surrounded by a system of rays. With some careful inspection, you will see that some of these rays stretch all the way to the opposite edge of the Moon. When a comet or an asteroid formed this crater, possibly 100 million years ago, it gouged out a piece of Moon

GALILEO AND THE FIRST CLOSEUP VIEWS OF THE MOON

Galileo was late hearing about the new device that could make far-off objects appear close. In September 1608, the Dutch spectacle maker Hans Lippershey had applied for a patent, but others were quick to apply for patents too. Before Galileo finally heard of the telescope in the summer of 1609, small 4x magnification telescopes were for sale in France.

Thus, Galileo did not invent the telescope. But he did increase its power to 9 times and, by November 1609, to 20. With this telescope he began his study of the Moon, recording its appearance as it went through a full cycle of phases. Amazed by the unevenness of its surface, Galileo made at least eight careful drawings showing mountains, plains, and many circular features.

In viewing the Moon's surface as something like the Earth's, Galileo had a basis for visualizing the Moon as a world like our own.

GALILEO GALILEI *as portrayed in 1637 by Justus Sustermans (1597–1681), and some of Galileo's drawings of the Moon.*

some 50 miles (90 km) across, sending rocks streaking across the Moon to form these rays. All craters have ray systems when they are young, but the loose material does not survive the billions of years of micro-meteorite hits and other subtle effects that erode the surface.

The rays stand out when the Moon is full, but at other times they cannot be seen. Instead, you will see a twilight zone between its bright and dark portions. Known as the terminator, this is a region of changing shadows, where craters and mountain ranges stand out in stark relief. Regular observers of the Moon study the same area in many different lighting conditions in order to appreciate the tremendous detail laid out before them by even the smallest telescope.

As far as photography is concerned, the Moon is bright enough for exposure times of 1 second or less, which makes it an ideal subject for the novice astrophotographer.

TIDES

Back on Earth, anyone who has visited an ocean beach will have been aware of the rise and fall of the tides. In many places in the world, the landscape changes radically as the tidal currents surge back and forth, and they play a pivotal part in the rich natural life of the shoreline.

The Moon's gravity, and to a lesser extent that of the Sun, causes two high tides on the Earth each day—one every 12 hours and 25 minutes. This interval of time reflects the Earth's rotation and the Moon's revolution around the Earth.

But why are there two tides a day? The gravitational pull of the Moon tugs at the Earth, causing the waters on the side facing it to pile up, accounting for one high tide. The high tide on the other side of the Earth arises because the gravitational pull of the Moon is such that the Earth itself is pulled a little toward the Moon. This movement results in the waters on the far side being "left behind" and piling up. In between these high tide regions are the regions where the water level is at its lowest —the low tides. (See illustration on p. 92)

(See illustration on p. 92)

FACT FILE

Distance from Earth: 239,000 miles (384,000 km)

Sidereal revolution period (about Earth): 27.3 days

Mass (Earth = 1): 0.012

Radius at equator (Earth = 1): 0.272

Apparent size: 31 arc minutes

Sidereal rotation period (at equator): 27.3 days

Everyone is a moon and has a dark side which he never shows to anybody.

Pudd'nhead Wilson's New Calendar, MARK TWAIN (1835–1910), American writer

MARE CRISIUM *is the prominent, dark, circular "sea" in this view from the Apollo 8 spacecraft.*

MERCURY: PLANET *of* EXTREMES

Even though Mercury is the planet that orbits closest to the Sun, it has one of the coldest nights in the Solar System.

Mercury is the most difficult of the naked eye planets to recognize. Moving swiftly through the sky, it is never more than 28 degrees from the overwhelming brilliance of the Sun. It is visible for a period of weeks in the evening sky and, much later, reappears in the morning sky.

MESSENGER OF THE GODS
Mercury, portrayed with a winged cap, in an eighteenth-century oil painting by François Boucher (1703–1770).

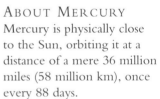

Although the Chinese and the Egyptians knew of this planet, it is the Roman name for it that we use today. Mercury was an ancient Roman deity—the god of commerce who is commonly identified with the Greek god Hermes, the nimble messenger of the gods, with his winged cap or sandals.

ABOUT MERCURY

Mercury is physically close to the Sun, orbiting it at a distance of a mere 36 million miles (58 million km), once every 88 days.

Although all the old astronomy texts tell us that Mercury's day is the same length as its year, radar studies made in 1965 proved this to be wrong. In fact, the planet rotates once every 59 days. Since the planet's day is two-thirds the length of its year, from some places on Mercury one would be able to watch the Sun zigzagging in a crazy path across the sky, staying above the horizon for 90 days at a time.

Mercury is without a protective atmosphere to shield its surface from the intense barrage of solar radiation it receives and to smooth out the variation in temperature between day

THE CRATERS OF MERCURY
are graphically revealed in this mosaic of images taken by the Mariner 10 spacecraft in 1974, from a distance of 124,000 miles (200,000 km).

and night. Its sunny-side temperature is as much as 750 degrees Fahrenheit (400 degrees Celsius), and then it plummets to -330 degrees Fahrenheit (-200 degrees Celsius) at night-time.

In 1974 and '75, NASA's Mariner 10 spacecraft made three passes of the planet and took a series of outstanding photographs that showed Mercury's surface to be pockmarked with craters, large and small. Like the Moon, Mercury still carries a dramatic record of the heavy bombardment with debris it received during the Solar System's early history.

MERCURY IN TRANSIT *Tiny Mercury, dwarfed by sunspots (center), cutting across the edge of the Sun's disk in 1993*

OBSERVING MERCURY

Surprisingly, many astronomers have never seen Mercury. If you know where and when to look, however, it is an easy planet to spot, either low in the west after the sun has gone down or in the east round about dawn. Through a small telescope you will be able to make out its phase as it moves around the Sun, mimicking the Moon as it shrinks from gibbous to a thin crescent. You will find that frequently the seeing is poor close to the horizon, so you will have to be satisfied with a colorful, dancing image.

At irregular intervals, about a decade apart, Mercury passes directly between the Earth and the Sun. These events are known as transits. Using safe techniques (see Viewing the Sun Safely, p. 65), Mercury can be seen as a tiny black dot, slowly making its way across the face of the Sun.

FACT FILE

Distance from Sun:
0.39 AU

Sidereal revolution period (about Sun): 88.0 days

Mass (Earth = 1): 0.055

Radius at equator (Earth = 1): 0.38

Apparent size: 5–13 arc seconds

Sidereal rotation period (at equator): 58.7 days

Moons: none

EVENING STARS *The Moon, with Mercury just above, and brilliant Venus and stars of Scorpius higher in the sky*

OBSERVING A TRANSIT OF MERCURY *using a telescope with a full aperture solar filter*

VENUS: *the* EVENING *and* MORNING STAR

Venus, shrouded in cloud, is the brightest of the planets in our sky, but she reveals her secrets reluctantly.

THE BIRTH OF VENUS *A detail from the famous painting by Sandro Botticelli, completed in the 1470s, which hangs in the Uffizi Gallery in Florence*

Once called Hesperus, when it was a morning star, and Phosphorus, as an evening star, Venus is the name of the ancient Roman goddess of beauty and love. The planet is somewhat misnamed, for even though it is about the same size as the Earth, conditions there are as close to an inferno as one can imagine.

Venus is covered by clouds of water vapor and sulfuric acid so dense that we cannot see its surface without using advanced radar systems like the one on NASA's Magellan spacecraft. Temperatures on the planet's surface approach 860° Fahrenheit (460° Celsius), the hottest in the Solar System, in a pressure-cooker environment where a barometer reading would be a hundred times higher than on Earth.

Since the atmosphere is almost entirely carbon dioxide, we can conclude that Venus suffers from a greenhouse effect gone wild. Radiation from the Sun heats the surface, just as it does Earth, but the heat cannot escape through the thick cocoon of carbon dioxide and clouds. Even at night, the temperature hardly drops at all.

Ground views from Venera landers revealed a desolate landscape with slabs of volcanic rock on coarse-grained dirt. In 1993, the Magellan spacecraft completed the most comprehensive sequence of maps of Venus's surface, revealing details of continent-like features and craters. The low number

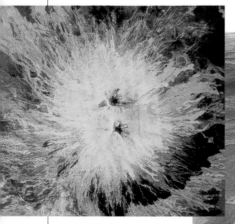

VOLCANOES *(above) Radar imaging from orbit revealed volcanoes that computers can display from any angle. The view of Maat Mons (right) shows lava flows extending hundreds of miles from the 3 mile (5 km) high peak.*

PHASES OF VENUS *(right) Venus (and Mercury) undergo phases like the Moon because their orbits are closer to the Sun than Earth's. The planet is at its brightest when showing only a large, thin crescent because it is then very close to the Earth.*

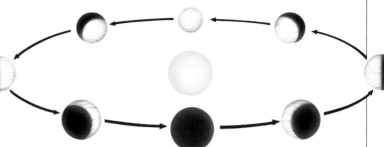

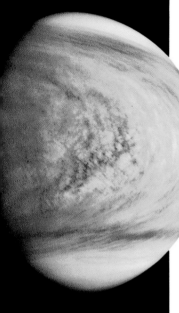

THE CLOUDS OF VENUS *revealed some details to imaging by the Pioneer Venus orbiter in ultraviolet light. Visible light photographs show almost nothing of the clouds, which are made of sulfuric acid and lie 40 miles (60 km) above the surface.*

Though the day still lingers, the rose-scattering fire of the evening star already scintillates through the azure sky.

WILLEM KLOOS (1859–1938), Dutch poet and essayist

OBSERVING VENUS

You cannot miss Venus when it is well placed in the evening sky, being by far the brightest "star" of all. In fact, if you know exactly where to look, you can even see Venus in broad daylight. As it rushes around the Sun, Venus spends about one season in the evening sky, followed by another in the morning sky, ranging up to 48 degrees away from the Sun.

Through a small telescope, Venus is blindingly bright, and, like the Moon and Mercury, it has phases. When it presents a crescent phase, you may see a faint glow on the darkened region, known as the Ashen Light, which varies in intensity over time. Other details of the clouds are subtle and ephemeral, making the planet brilliant overall, but a little bland.

of craters resulting from impacts by asteroids and comets suggests that the surface is relatively young— perhaps the planet was "resurfaced" by volcanic activity as recently as 800 million years ago.

The many features that become visible from detailed mapping need to be named for ease of identification. The International Astronomical Union (IAU) decided that the names given to all features on Venus should (appropriately enough) be female. Thus the Greek goddess Aphrodite's name has been given to a "continent" and a crater carries the name of jazz singer Billie Holiday. There is one exception, though: Scottish physicist James Clerk Maxwell is "the only man on Venus", with Maxwell Montes, the Maxwell Mountains, named after him.

FACT FILE

Distance from Sun: 0.72 AU

Sidereal revolution period (about Sun): 225 days

Mass (Earth = 1): 0.81

Radius at equator (Earth = 1): 0.95

Apparent size: 10–64 arc seconds

Sidereal rotation period (at equator): 243 days

Moons: none

BRILLIANT VENUS *is a feature of the western sky after sunset for several months each year. Here, we see Venus accompanied by the crescent Moon.*

EARTH: *a* LIVING WORLD

The Earth is unique in the Sun's family—a fragile sphere
cloaked in water and oxygen, with a
wonderful diversity of life.

Among all the planets, the Earth is unique in several respects. To an observer elsewhere in the solar system, it displays a dynamic atmosphere that is remarkable in having a 21 percent oxygen content. Clouds of water vapor obscure a variable portion of the surface, but make it a brilliant beacon in the inner solar system. Perhaps most remarkable of all is the fact that over

THE CREATION

God creates Adam in a detail from Michelangelo's fresco in the Sistine Chapel, Rome, commissioned 1508–12

70 percent of its surface is covered by water in either liquid or solid (ice) form. Liquid water is not seen on the surface of any other planet in the Sun's domain.

EARTHRISE FROM THE MOON

as seen during the first historic lunar landing by Apollo 11 in 1969

LIFE ON EARTH *The abundance and diversity of plant and animal life on Earth is nowhere more evident than in the rainforests of the tropical regions of the planet. Here, a female orangutan and her infant in a rainforest in Borneo are representative of the more complex and intelligent creatures to which Earth is home.*

EARTH FROM APOLLO 17 *A fantastic view of our planet that extends all the way from Saudi Arabia south to the Antarctic icecap.*

GEOLOGICAL ACTIVITY

Geologically, the Earth is very active, in stark contrast with its sister planet, Venus. The Earth's surface is split into plates, floating on a rocky mantle. Earthquakes, volcanic activity, and mountain-building are concentrated along the boundaries of these plates.

As a result of this geological activity, plus weathering by wind and rain, the Earth's surface does not show many signs of the pummelling by impacts that are so common elsewhere in the solar system. The craters are, for the most part, worn away or hidden.

THE PRESENCE OF LIFE

Most extraordinary of all, so far as we have been able to determine, the Earth is the only planet in the solar system that sustains life.

In fact, the presence of life on Earth is not immediately obvious from space. Perhaps the huge fields of the grain-growing regions of the world would provide evidence of farming. Certainly, the Earth's night-side would be distinguished by the carpets of lights marking the cities. In recent years, another indication of the presence of intelligent life would have been provided by radio emissions leaking into space.

Someday I would like to stand on the moon, look down through a quarter of a million miles of space and say, "There certainly is a beautiful earth out tonight."

LT. COL. WILLIAM H. RANKIN
(b. 1920), American author

FACT FILE

Distance from Sun:
1.00 AU

Sidereal revolution period (about Sun): 365.26 days

Mass (Earth = 1): 1.0

Radius at equator (Earth = 1): 1.0

Sidereal rotation period (at equator): 23.9 hours

Moons: 1

MARS: *the* RED PLANET

Mars is well embedded in the popular imagination as the most likely abode of life elsewhere in the Solar System.

In 1938, Orson Welles's adaptation of H.G. Wells's book, *War of the Worlds,* terrified many radio listeners into believing that Martians had really invaded New Jersey.

The basis for this story was set late in the nineteenth century by the Italian astronomer Schiaparelli, who recorded long, straight lines that seemed to criss-cross the surface of Mars. He called these *canali*, meaning channels, implying natural features in a landscape; but in English the name became translated as canals, which gave the sense of artificially made structures.

MARS, GOD OF WAR, *in a detail from a dramatic seventeenth-century painting by Giovanni Battista Carlone*

MARTIAN LIFE?

Percival Lowell became fascinated with these canals, which he supposed to be a complex network that a civilization of intelligent Martians had built to move scarce water from one place to another.

Other observers, using equally powerful telescopes, were unable to see any canals, but the possibility of life on Mars persisted in the popular imagination. The idea of Martian canals was finally laid to rest in July 1965 when the American spacecraft Mariner 4 passed the planet and found no signs of construction at all. In 1976, two Viking landers probed the surface of Mars but could find no evidence of life in the cold, desert-like environment.

MARS FROM VIKING, *with the thin south polar cap visible near the bottom*

ABOUT MARS

Mars has a cratered surface, but its craters are not as thickly clustered as those on the Moon

THE SURFACE OF MARS *at the Viking lander sites featured fine red sand strewn with rocks up to several feet (around a meter) across.*

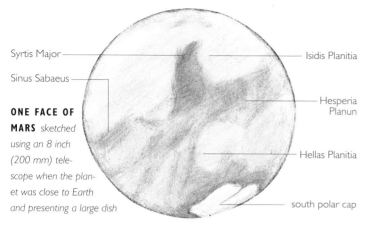

Syrtis Major

Sinus Sabaeus

Isidis Planitia

Hesperia Planun

Hellas Planitia

south polar cap

ONE FACE OF MARS *sketched using an 8 inch (200 mm) telescope when the planet was close to Earth and presenting a large dish*

"The chances against anything man-like on Mars are a million to one," he said.

War of the Worlds,
H.G. WELLS (1866–1946),
English writer

or Mercury because weathering, largely from fierce windstorms, has worn them away. There is also tantalizing evidence of erosion by flowing water some time in the past. Where is the water now? A little is locked in thin polar caps of water and carbon dioxide ices. Perhaps the rest lies in permafrost beneath the surface.

Mars has some spectacular features, one of the most prominent being Olympus Mons, an enormous volcano, seemingly dormant, which is larger than any mountain on Earth. The Valles Marineris is equally remarkable—a system of canyons up to 4 miles (7 km) deep, forming a great gash stretching 2,500 miles (4,000 km) across the planet.

OBSERVING MARS

Being farther from the Sun than Earth, Mars can appear at any place in the sky on the ecliptic, rather than always staying near the Sun, like Mercury and Venus. When closest to the Earth, it can

be a mere 35 million miles (56 million km) from us and twice as bright as Sirius, the brightest star. At other times, the eccentricity of its orbit may place it about 250 million miles (400 million km) away. This far from Earth it looks tiny even through small telescopes.

When you can see Mars clearly, its details are astonishing. Like Earth, it has seasons, and you may be able to see one of the polar caps grow or shink as a season progresses. The dark surface features in its southern hemisphere, like Solis Lacus (the "Eye of Mars") and Syrtis Major, are usually easy to see if Mars's rotation brings them into view. Other features, like the northern hemisphere's Mare Acidalium, vary in intensity from year to year. In fact, most Martian features change somewhat from one viewing to another. Thin clouds often form around the mountains, and if the planet has just passed its perihelion— the closest point in its orbit to the Sun—it might suffer a planet-wide dust storm that will obscure almost all its features for several weeks.

PERCIVAL LOWELL *is famous for his belief in life on Mars, but his search for a planet beyond Neptune was of greater significance. This planet, Pluto, was finally discovered in 1930, 14 years after his death.*

MARS'S MOONS

In 1726, Jonathan Swift wrote in *Gulliver's Travels* of the discovery of two Martian moons. It would be a century and a half before Asaph Hall actually found them! These tiny moons, which became known as Phobos and Deimos, are probably small asteroids, captured into orbit by Mars's gravitational pull.

MARTIANS *Aliens from Mars have stereotypically been depicted as odd-looking green creatures. This character appeared on the cover of Astounding Science Fiction in 1954.*

FACT FILE

Distance from Sun:
1.52 AU

Sidereal revolution period (about Sun): 687 days

Mass (Earth = 1): 0.11

Radius at equator (Earth = 1): 0.53

Apparent size: 4–25 arc seconds

Sidereal rotation period (at equator): 24.6 hours

Moons: 2

JUPITER: *the* GREAT GIANT

Like its namesake, the supreme deity of the Romans, Jupiter is the giant of all the planets.

With over 300 times the mass of the Earth, and about 2½ times the mass of all the other planets put together, Jupiter is the dominant planet in the Solar System. It was the first planet studied by Galileo through his telescope.

ABOUT JUPITER

Together with Saturn, Uranus, and Neptune, Jupiter is a gas giant, far more massive and far less dense than the small, mostly rocky planets of the inner Solar System. Its atmosphere is a mixture of hydrogen, helium, methane, and ammonia. Beneath the cloud tops that we see lie layers of increasingly dense gases, with a small, rocky core in the middle.

Jupiter rotates fast—once around in under 10 hours! This flattens the planet's disk at the poles and drives dynamic weather patterns in the clouds which envelop the planet, causing its features to change rapidly. Its cloudy disk is banded with belts interspersed with bright but variable zones.

A WHIRLPOOL OF WIND

The largest storm on any planet is Jupiter's Great Red Spot— 30,000 miles (50,000 km) long and one-third as wide. It is the main feature of the clouds

GANYMEDE, a mortal youth abducted by Jupiter (disguised as an eagle), gave his name to the largest of Jupiter's moons.

of the Southern Hemisphere and is about four times larger than the Earth!

First seen in 1664 by Robert Hooke, it varies in intensity and tint from year to year. This huge whirlpool of wind is the most prominent example of similar ovals seen in the atmospheres of all the giant planets.

THE GALILEAN MOONS

Discovered by Galileo, Jupiter's four big moons are among the most fascinating in the Solar System, as revealed when Voyager 1 sped by in 1978.

Io, the closest of the four moons to Jupiter, is so affected by the planet's gravitational pull that it is in turmoil, its surface constantly being repaved by sulfurous volcanic eruptions. Further from Jupiter is Europa, whose icy surface, while also relatively fresh, resembles a planet-wide skating rink. The outer moons, Ganymede and Callisto, have older surfaces of ice, both scarred with impact craters, but differing in other features, and telling of quite different histories.

JUPITER AND THE RED SPOT *This color-enhanced image from Voyager 1 shows tremendous detail in the clouds, especially where the Great Red Spot carves its path.*

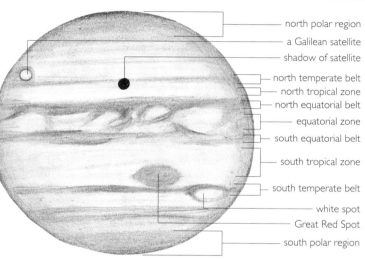

JUPITER AND THE GALILEAN MOONS *is an ever-changing view revealed by the smallest telescope.*

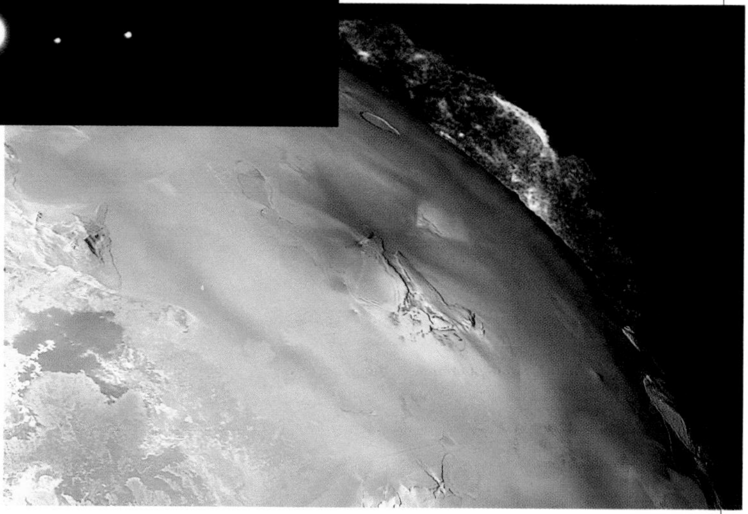

IO'S VOLCANIC ACTIVITY *is seen here producing plumes of material hanging in the sky above its disk.*

OBSERVING JUPITER

The easiest way to observe Jupiter is to repeat Galileo's experiment and keep a nightly log of the changing positions of the four large moons. On one night, all four moons may be visible, while on another you may see only two, with a third in front of Jupiter, betrayed only by its shadow on the planet.

Or you might find it interesting to draw Jupiter itself. With a small telescope, use just enough magnification to see the belts and zones clearly. Before you begin, look at the planet for a few minutes. Once you start (using a 2B pencil), you will need to outline the major features in about 10 minutes, since Jupiter's rotation carries features along quickly. You can fill in the detail later.

Labels on sketch:
- north polar region
- a Galilean satellite
- shadow of satellite
- north temperate belt
- north tropical zone
- north equatorial belt
- equatorial zone
- south equatorial belt
- south tropical zone
- south temperate belt
- white spot
- Great Red Spot
- south polar region

SKETCH OF JUPITER *made while observing with an 8 inch (200 mm) telescope. The unusual feature here is the Galilean satellite, and its shadow, captured in transit across the disk.*

FACT FILE

Distance from Sun:
5.20 AU

Sidereal revolution period (about Sun): 11.9 years

Mass (Earth = 1): 318

Radius at equator (Earth = 1): 11.2

Apparent size: 31–48 arc seconds

Sidereal rotation period (at equator): 9.84 hours

Moons: 16

THE GREAT RED SPOT *moves relative to the surrounding clouds, creating complex wave patterns which are revealed by color enhancement in this Voyager 1 image.*

249

SATURN: LORD *of the* RINGS

With its graceful system of rings,

Saturn is one of the sky's most rewarding

sights through any size of telescope.

SATURN, *god of the harvest, surrounded by symbols of Rome and agricultural produce*

S aturn is named after the ancient Roman god of the harvest, who was the father of Jupiter. When you step up to your telescope and see the small, distant globe of Saturn, surrounded by its exquisite rings, it will take your breath away.

The rings are composed of innumerable icy chunks that range in size from fine dust to house-sized blocks. They form the three broad rings visible from Earth, lettered from the outermost A, B, and C. The two Voyager spacecraft recorded hundreds of very

SATURN *seen with similar color enhancement to the image of Jupiter (p. 248). Clearly, the clouds of Saturn lack the banding and colors of Jupiter's atmosphere.*

narrow "ringlets" making up the rings. Between the bright A and B rings lies the Cassini Division, which appears to be a gap but was revealed by Voyager to be merely less densely packed with ringlets.

The likely origin of the rings was first proposed in the nineteenth century, when Edouard Roche, a French mathematician, suggested that if an object such as a

small moon approached too close to a planet, it could be torn apart by the planet's tidal force. Sometime in the past, one or several objects must have passed within Saturn's "Roche limit" and broken apart to form the rings.

We now know that Jupiter, Uranus, and Neptune all possess rings, but none of them are as grand or as easy to see as those of Saturn.

SATURN'S MOONS

Saturn has 18 known natural satellites, of which Titan is the second largest in

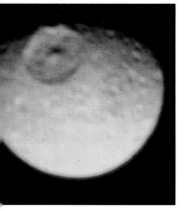

MIMAS *is only 240 miles (390 km) across, and this huge crater spans one-fourth of that!*

RINGS *The thousands of ringlets that make up Saturn's distinctive rings are not easy to see here, but this highly enhanced colour image reveals subtle variations in their composition.*

the Solar System, after Jupiter's Ganymede. Almost 3,200 miles (5,150 km) across, Titan has an atmosphere consisting of nitrogen and methane. It is the only satellite known to have an atmosphere.

Saturn's other satellites have icy surfaces and are heavily

CHANGING VIEWS *of Saturn's rings as it moves around the Sun result from the tilt of the planet's rotation axis.*

cratered, each with a distinct character. Mimas has a huge impact crater named Herschel, the result of a collision that must have almost torn the moon apart. Iapetus is a 900 mile (1,440 km) diameter enigma. Its trailing hemisphere, which faces where the satellite has just traveled in its orbit, resembles those of other satellites, but the leading hemi-sphere is one-tenth as bright —coated with a dark reddish layer of unknown origin.

OBSERVING SATURN

For first-time viewers, Saturn is a memorable sight in a telescope. Even through bino-culars, where the rings are not distinguishable, its elongated shape is unmistakable.

Although Saturn has more subtle shadings and shows less rapid changes than Jupiter, Saturn's rings are always fascinating. Almost any small telescope will show Titan, and at least three other moons will be apparent in a 6 inch (150 mm) telescope.

Unlike the moons of Jupiter, which seem to always be in a line, Saturn's moons

can be above or below the planet because, like the Earth, the planet is tipped relative to the plane of the Solar System.

As Saturn orbits the Sun once every 29 years, the aspect of the rings changes slowly. Some years they will be majestically spread out, and about once every 14 years they will be virtually edge-on and therefore almost invisible. Twice during each orbit the rings reach a maximum inclin-ation to the line of sight— once when they are visible from above and once when visible from below.

FACT FILE

Distance from Sun: 9.54 AU

Sidereal revolution period: 29.5 years (about Sun)

Mass (Earth = 1): 95.2

Radius at equator (Earth = 1): 9.5

Apparent size: 15–21 arc seconds (planet's disk)

Sidereal rotation period (at equator): 10.2 hours

Moons: 18

URANUS: PLANET UPENDED

Rolling along on its side, Uranus was

the first planet to be

discovered in modern times.

THE SKY GOD *Uranus, of Greek mythology, being overthrown by his Titan son, Cronus*

Tilted at an angle of 98 degrees to its plane of orbit, this giant green-blue planet lies on its side, which produces some strange effects. For example, during its 84-year orbit around the Sun, each of its poles periodically faces the Sun and experiences a day that lasts 42 years.

A gas giant like Jupiter and Saturn, Uranus has a seemingly impenetrable atmosphere of hydrogen, helium, methane, and ammonia. The temperature in its upper clouds is

HERSCHEL'S DISCOVERY *as he recorded it himself on March 13, 1781*

estimated to be about –330 degrees Fahrenheit (–200 degrees Celsius), a temperature at which the ammonia would exist in the form of ice crystals.

URANUS'S MOONS

The main surprises of Voyager 2's visit to Uranus were associated with its moons, of which the largest, Titania, is about half the size of our Moon.

Except for Umbriel (the "melancholy sprite" in Pope's poem *The Rape of the Lock*), all five major satellites are named after Shakespearean characters.

VOYAGER AT URANUS *A montage of Voyager images of moons surrounding Uranus: (clockwise from bottom) Ariel, Umbriel, Oberon, Titania, and Miranda*

Titania, Oberon, and Umbriel appear to be relatively quiet geologically, but Miranda and Ariel have surface features that suggest violent activity in the past.

In 1977, the passage of Uranus in front of a star was observed from Earth and led to the discovery of nine faint, narrow rings around the planet. Two more were found by Voyager, plus many tiny ringlets and arcs of rings.

OBSERVING URANUS

Viewed through a telescope, Uranus appears as a greenish disk, never brighter than about 6th magnitude, and it is slightly elliptical because of its rapid rotation. The combination of a small disk and seeing effects caused by the Earth's atmosphere make it difficult to see the planet in any great detail. Even the Voyager images showed a largely featureless atmosphere.

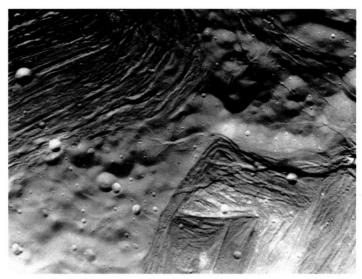

MIRANDA has fracture patterns and dramatic landscape features that might indicate it was once shattered and then reassembled after some collision early in its history.

FACT FILE

Distance from Sun:
19.2 AU

Sidereal revolution period (about Sun): 84.0 years

Mass (Earth = 1): 14.6

Radius at equator (Earth = 1): 4.0

Apparent size: 3–4 arc seconds

Sidereal rotation period (at equator): 17.9 hours

Moons: 15

WILLIAM HERSCHEL AND THE DISCOVERY OF URANUS

Born in Hanover, Germany, in 1738, William Herschel was the son of a musician, and became a skilled musician himself. When he moved to England in 1757, he aimed to establish his reputation in the world of music, but he became interested in skywatching after buying a book on astronomy.

By 1781, Herschel was involved in a systematic survey of the sky, using a telescope he had built himself. On March 13 that year, he found a star in Gemini that, he wrote, "appeared visibly larger than the rest...I suspected it to be a comet." By the end of August, with Gemini rising in the morning sky and the new object visible once more, Finnish mathematician Anders Lexell announced that Herschel had found a new planet.

Herschel's career blossomed. In 1782, George III appointed him King's Astronomer, enabling Herschel to devote his life to astronomy. He made ever larger telescopes and discovered two new satellites of Saturn, and two satellites of Uranus.

RINGS Voyager 2 looked back toward the Sun to capture this view of the rings around Uranus. Star images in the background trailed during the exposure.

253

BLUE NEPTUNE

This planet is the last and most distant of the giant outer worlds revealed by Voyager 2.

Thonis planet is named after the Roman god of water, in earlier times also a god of fertility. This relationship with the sea has turned out to be highly appropriate, as photographs of the planet show it to be a deep, crisp blue. On August 24 and 25, 1989, Voyager 2 flew by Neptune and its moons. Its pictures showed that Neptune's atmosphere has zones like Jupiter, and a giant storm now called the Great Dark Spot. Images of the planet also showed bright clouds of methane ice crystals floating in the atmosphere and confirmed that the planet has a faint ring system. This had been suspected by astronomers for several years.

NEPTUNE'S MOONS

Only a month after Neptune was discovered, William Lassell discovered Triton, a satellite of Neptune close in size to our Moon. Unlike all other large planetary satellites, Triton swings round Neptune in a retrograde orbit—in other words, it moves in the opposite direction to the planet's rotation. A smaller satellite, Nereid, discovered in 1949, has a diameter of about 200 miles (320 km) and orbits in a long ellipse.

Voyager 2 located six new moons in between Neptune and Triton, of which Naiad is fairly large, at some 250 miles (400 km) across.

NEPTUNE *(right) presents a deep blue disk with bright clouds of methane ice crystals. The Great Dark Spot is the size of the Earth.*

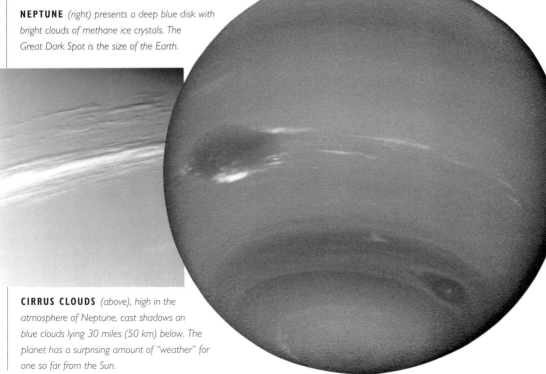

CIRRUS CLOUDS *(above), high in the atmosphere of Neptune, cast shadows on blue clouds lying 30 miles (50 km) below. The planet has a surprising amount of "weather" for one so far from the Sun.*

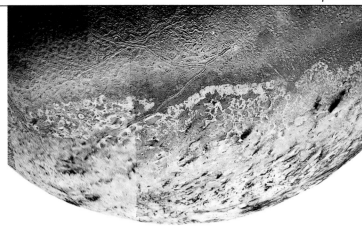

Is this ... A time to sicken and to swoon, When Science reaches forth her arms, To feel from world to world, and charms her secret from the latest moon?

In Memoriam (published in 1850, four years after Neptune and its moon, Triton, were discovered), ALFRED LORD TENNYSON (1809–92), English poet

OBSERVING NEPTUNE

To the casual observer, Neptune may be something of a challenge to find, presenting a tiny bluish disk of about 8th magnitude.

You will need accurate details of its position and good charts, perhaps from a current astronomy magazine, to pick it out from a myriad of faint stars. The thrill is definitely in the hunt!

TRITON'S SOUTH POLE *has dark streaks which may be small volcanoes or geysers driven by liquid nitrogen.*

FACT FILE

Distance from Sun: 30.0 AU

Sidereal revolution period (about Sun): 165 years

Mass (Earth = 1): 17.1

Radius at equator (Earth = 1): 3.88

Apparent size: 2.5 arc seconds

Sidereal rotation period (at equator): 19.2 hours

Moons: 8

J.C. ADAMS *(below, left) His involvement with the discovery of Neptune was the first step in a distinguished career in mathematics and astronomy.*

LE VERRIER *is shown here working out the position of Neptune, in this artist's impression from Camille Flammarion's* Astronomie Populaire *(1881)*

ADAMS, LE VERRIER, AND THE DISCOVERY OF NEPTUNE

In 1845, a 23-year-old student named John Couch Adams completed calculations pinpointing a new planet that he believed was affecting the orbit of Uranus. In September of that year, he sent his calculations to his professor, John Challis, who gave them to George Airy, England's Astronomer Royal. Airy seemed interested, but did not order a search.

At the end of 1845, a French astronomer named Urbain Jean Joseph Le Verrier published his prediction of the position of a new planet, within one degree of where Adams had said it would be. On July 9, 1846, Airy suggested that Challis search for the object, so Challis mounted a star-by-star search over a large area of sky around the predicted position, going over the new planet twice without recognizing it.

In the meantime, Le Verrier, having failed to instigate a search at the Paris Observatory, approached Johann Galle at the Berlin Observatory. Intrigued, Galle began a search at Le Verrier's predicted position with his assistant Heinrich d'Arrest. It did not take the two long to find an eighth magnitude "star" that did not belong. As the object moved among the stars, it was confirmed as the new planet.

After the discovery was announced, Airy tried to have Adams credited with the discovery as well, which infuriated the French. Adams and Le Verrier kept apart from the controversy and when they met, some time later, they became close friends.

PLUTO *and* BEYOND

The Solar System's most distant planet,
Pluto loops around the Sun in a strange orbit,
accompanied by its unusually large moon, Charon.

PLUTO, *also called Hades, ruled the kingdom of the underworld with his queen Persephone. Both appear in this sculpture by Bernini (1598–1680).*

Pluto is named after the Greek god of the underworld. Lowell Observatory astronomers came up with the name shortly after its discovery, following a suggestion made by Venetia Burney, an 11-year-old English girl, and it appropriately commemorates Percival Lowell in its first two letters.

In 1978, James Christy discovered that Pluto has a moon, now named Charon. A fortuitous series of events, when Pluto and Charon repeatedly eclipsed each other in the late 1980s, allowed observers to determine that Pluto's diameter is only 1,430 miles (2,300 km). Charon is fully half as big. Observations of Pluto as it occulted (covered) a star revealed that the planet has a thin atmosphere that is likely to condense on the surface as frost as Pluto moves farther from the Sun in its very eccentric orbit.

Clearly, Pluto does not have the mass to influence the orbits of Neptune and Uranus in the

A LARGE SATELLITE *This Hubble Space Telescope photograph was taken before the 1994 repair mission. The separation between Pluto and Charon is about 12,000 miles (20,000 km).*

FACT FILE

Distance from Sun:
39.5 AU

Sidereal revolution period (about Sun): 249 years

Mass (Earth = 1): 0.002

Radius at equator (Earth = 1): 0.18

Apparent size: 0.04 arc seconds

Sidereal rotation period (at equator): 6.39 days

Moons: 1

PLUTO AND CHARON *An artist's conception of ice-covered Pluto (foreground) and Charon, its satellite. The distant Sun casts a cold light on these far-off worlds.*

way Lowell had expected. Its discovery was a triumph not of mathematical prediction but of Tombaugh's thorough search and keen eye. Perhaps better early observations would have removed the apparent anomalies in the motions of Uranus and Neptune which formed the basis of the hunt for another planet.

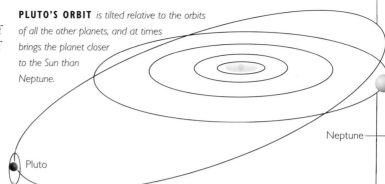

PLUTO'S ORBIT *is tilted relative to the orbits of all the other planets, and at times brings the planet closer to the Sun than Neptune.*

Neptune

Pluto

OBSERVING PLUTO

Pluto is not a target for the beginning skywatcher. At about magnitude 14, it is perhaps just within the range of an 8 inch (200 mm) telescope, but it must be identified among the myriad of faint stars by watching its slow movement among them over several nights.

PLUTO, *Walt Disney's lovable droopy-eared dog, made his first appearance a few months after the discovery of his celestial cousin.*

ANOTHER SOLAR SYSTEM? *Beta Pictoris is about 50 light years away and is surrounded by a disk of icy and rocky material, revealed here by blocking light from the star itself.*

ASTEROIDS IN THE OUTER SOLAR SYSTEM

In 1977, Charles Kowal discovered the asteroid 2060 Chiron orbiting in an elliptical orbit between Saturn and Uranus, the first asteroid seen in the outer Solar System. In 1988, Chiron surprised astronomers by doubling in brightness and subsequently showed a faint, comet-like coma. With an estimated diameter of about 125 miles (200 km), Chiron may represent a class of large asteroids or comet nuclei lurking undetected in the outer Solar System. A second object, 5145 Pholus, has since been discovered. In 1992 David Jewitt and Jane Luu of the University of Hawaii found 1992 QB1, the first example of an object of similar size orbiting the Sun beyond the outermost planets, Neptune and Pluto.

CLYDE TOMBAUGH AND THE DISCOVERY OF PLUTO

Clyde William Tombaugh (shown below with a telescope he designed to photograph satellites) was born in Streator, Illinois, in 1906. Interested in astronomy from the time he was a boy, he made his own telescope and used it to sketch the planets. He sent drawings of Mars and Jupiter to the Lowell Observatory to get their opinion of his work. The director, V. M. Slipher (of red shift fame; see p. 50) must have been impressed, for he offered Tombaugh a job at Lowell as observer.

After arriving at Lowell, Slipher told Tombaugh that his main project would be to continue the search for a new planet that Percival Lowell had begun in 1905. The search method involved using a device called a blink comparator to examine two photographic plates of the same part of the sky taken a few nights apart. Stars would remain still, but any object that had moved during the interval would move back and forth.

On February 18, 1930, Tombaugh was comparing two plates centered on the bright star Delta (δ) Geminorum, not far from where Uranus had been discovered almost 150 years before. His sharp eyes picked out a faint moving object. After checking the image with a third plate he had taken, it was clear that he had found what he was looking for.

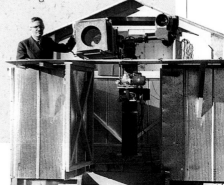

ASTEROIDS:
THE MINOR PLANETS

Rocky objects, smaller than planets,

that orbit the Sun, asteroids promise to tell us much

about the development of our Solar System.

In 1766, Johann Titius of Wittenberg, Germany, divided the distance between the Sun and Saturn, then the farthest known planet, into 100 units or "parts". Mercury is thus 4 parts from the Sun, Venus is 7 (or 4 + 3), Earth 10 (or 4 + 6), and Mars 16 (or 4 + 12). Titius discovered that if you keep doubling the second number, you reach a planet, right out to Saturn—except that there is no planet at 28.

Johann Bode, of the Berlin Observatory, was intrigued by Titius's theory and started a campaign to search for the missing planet beyond Mars. So enthusiastic was he that the theory is now often called Bode's Law rather than the Titius–Bode Law.

THE CELESTIAL POLICE

In 1800, a group calling themselves the Celestial Police came together in Germany to organize a systematic search for the missing planet. They divided the plane of the planets, the ecliptic, into several regions and assigned an astronomer to search each region.

On January 1, 1801, they were pre-empted by Giuseppe Piazzi of the Palermo Observatory in Sicily, who discovered a starlike object that behaved like a planet. Now called Ceres, this minor planet is in the correct place to satisfy Bode's Law, but, at less than 600 miles (1,000 km) across, it is rather small. On March 28,

ASTEROID 951 GASPRA *has its color variations exaggerated in this view from the Galileo spacecraft.*

ASTEROID 243 IDA *is 32 miles (52 km) long, more than twice the size of Gaspra.*

1802, the Celestial Police found a second asteroid, now known as Pallas, and later they found Juno and Vesta.

In more recent times, the pace of discovery has increased, and we now know the orbits of more than 15,000 asteroids, most of which fall into the gap of the Titius–Bode Law, but we still are not certain of the significance, if any, of this law.

TYPES OF ASTEROIDS
Main Belt Asteroids
The majority of asteroids with known orbits lie within the main belt of asteroids between the orbits of Mars and Jupiter. Scientists have recently been trying to piece their histories together by defining different families. Working back in time from their orbits, it is possible to deduce that, for each family, a single large asteroid broke up after a collision to produce the fragments we see today.

Trojans
There are two regions in Jupiter's orbit where asteroids can become trapped. Called Lagrangian Points, one is

ELEANOR HELIN

One of the most successful asteroid discoverers, Eleanor Helin began her work in 1969 by researching a dozen or so Apollo asteroids. These asteroids approach the Earth from time to time as they orbit the Sun, and are capable of hitting us. Using the 18 inch (450 mm) diameter Schmidt camera at Mount Palomar, Helin began a search for more of these asteroids. She recalls those difficult early days as the first woman observer working there regularly. "I'm sure Palomar didn't think I would last a year," she says.

Helin's search program is still going strong, and she has discovered numerous asteroids and comets. At the time of the Camp David Agreement, in 1978, Helin discovered a near-Earth asteroid. She subsequently named it Ra-Shalom, after the Egyptian Sun god Ra and the Hebrew word for peace, to honor the peace accord between Israel and Egypt.

about one-sixth of an orbit in front of Jupiter and the other is one-sixth of the way behind it. Asteroids in these regions are called Trojans, and are traditionally named after figures from the legendary Trojan War. The Greeks (with 624 Hektor, a Trojan spy) precede Jupiter, and the Trojans follow it (along with Greek spy 617 Patroclus).

Near-Earth Asteroids
In 1989, Henry Holt discovered an Apollo asteroid that came within half a million miles (800,000 km) of Earth. In 1993, Tom Gehrels found another, only about 3 yards (3 m) wide, that had approached to within 90,000 miles (140,000 km) of Earth— well within the Moon's orbit.

We now know of about 200 larger asteroids that have orbits which cross the Earth's orbit, and all of them are capable of hitting the Earth some day. There are probably ten times that number, a mile or more across, that have yet to be found.

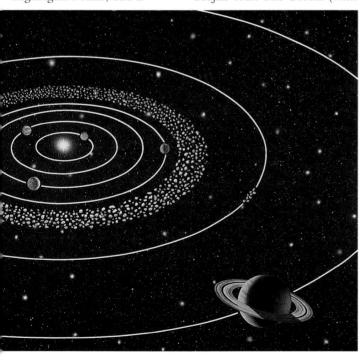

THE ASTEROID BELT, *and the Trojan asteroids which share the orbit of Jupiter. In reality, their positions are not this well localized.*

COMETS: *the* COSMIC WANDERERS

Streaking from the outer Solar System, comets no longer evoke the superstitious beliefs and terror they once did.

ADORATION OF THE MAGI *by Giotto di Bondone (1266–1336) shows a comet as the star of Bethlehem, perhaps inspired by the 1301 appearance of Halley's Comet.*

TWO TAILS *stream away from a comet when it is close to the Sun—a straight (blue) gas tail and a curving (yellow) dust tail.*

When a comet is far from the Sun, it is a cold body, perhaps a few miles across, that looks like a large, dirty snowball. We believe that most reside in a vast sphere of comets surrounding the Sun, known as the Oort Cloud, after the Dutch astronomer Jan Oort. This sphere lies well beyond the orbit of the most distant planets. Occasionally, one dirty snowball is perturbed onto a path inward toward the brilliance of the Sun. As it closes in, the ice begins to boil away, and a coma of gas and dust develops. The material leaves the comet to form separate gas and dust tails streaming away from the Sun. Sometimes a burst of material erupts in the form of a jet.

Occasionally a comet passes close to a planet, usually Jupiter, and the planet's gravity changes the comet's orbit. Repeated encounters may result in a new orbit that causes the comet to return over and over again to the inner Solar System. With its period of 76 years, Comet Halley is the best known example of such a periodic comet.

Old men and comets have been reverenced for the same reason, their long beards and pretences to foretell events.

JONATHAN SWIFT (1667–1745), Anglo-Irish writer

COMET WEST (1975a) *(right) showed a very faint blue gas tail pointing straight up and a more prominent yellow dust tail, as it rounded the Sun in 1976.*

CAROLINE HERSCHEL

As William Herschel's interest in astronomy grew (see p. 253), so did that of his younger sister, Caroline Lucretia. In addition to helping her brother with his work, in the 1780s she began hunting for comets using a 6 inch (150 mm) diameter reflector that he had built. On August 1, 1786, when William was away in Germany, she discovered her first comet.

Caroline Herschel's second comet find, at the end of 1788, turned out to be a periodic comet, returning about every 150 years. She discovered two comets in 1790 and another in late 1791. She found her eighth and last comet in 1797.

HALLEY'S COMET *was on the border of the constellations of Sagittarius and Capricornus and well placed for observers in the Southern Hemisphere when this photograph was taken in mid-March, 1986.*

sky is, you might find one of these fairly soon after you begin searching. If you do, plot its position carefully in a star atlas, or on the appropriate constellation chart in this book. If your chart does not show any fuzzy object there, wait a day and see if it has moved.

If you find a fuzzy object you think is a comet, be sure to confirm that it moves from one night to another. Make absolutely sure, perhaps by getting an experienced observer to view it. Then report it to the Central Bureau for Astronomical Telegrams (CBAT) (see Resources Directory). Make sure to provide an accurate position, a description, and an estimate of its brightness. If you report the comet before anyone else, it may be named after you!

COMET-FINDING POINTERS

Finding a comet generally takes a lot of time, experience, and patience. A dose of luck helps too! Since most new comets are faint, hunt only on a moonless night when the sky is clear and dark, using a telescope with a wide field of view.

Do your search in a pattern, moving your telescope slowly enough to see which stars and fuzzy objects are within each field of view.

The sky is full of fuzzy objects like galaxies, nebulae, and star clusters, many of which masquerade as comets. Depending on how dark the

TWO MODERN COMET FINDERS

At the end of 1973, Comet Kohoutek passed by the Sun, exciting many people the world over, but it proved not to be as bright as expected. Hardly anyone noticed a second comet, virtually as bright as Kohoutek, that rounded the Sun a few months later. Detected from Adelaide, Australia, this was to be William Bradfield's second comet discovery.

Since then, Bradfield has discovered more comets visually —that is, with his eye to the eyepiece of his telescope—than any other living person. During one intense period he found 2 comets within 10 days, but after 16 discoveries, he is still searching for more.

While William Bradfield hunts with his eye at the eyepiece, as do most amateur comet hunters, Carolyn Spellmann Shoemaker searches photographic films for comets and asteroids.

Carolyn Shoemaker began her career in astronomy by assisting her husband Eugene. By 1981, the Shoemakers were taking photographs at Mount Palomar's

18 inch (450 mm) telescope, and "eagle-eyed Carolyn," as Gene calls her, was starting to find new objects.

In 1983, Carolyn found her first comet. By 1991, she had found her twentieth, and by 1993 her thirtieth. Her name is now on more comet discoveries than anyone else's.

How Comets are Named

According to a tradition going back more than two centuries, new comets are named after their discoverers. Until a comet is confirmed, up to three independent observers can be credited with having found it.

The CBAT initially attaches a designation to a new comet. Periodic Comet Halley, 1982i, for example, was the ninth comet to be discovered or recovered in 1982 ("i" being the ninth letter of the alphabet). Some years after their original sighting, the CBAT gives comets designations based on

COMET SHOEMAKER-LEVY 9
Resembling a string of pearls or lights on a spaceship, Comet S-L 9 broke into fragments—each less than a mile or two across—when it made a close approach to Jupiter in 1992.

the order they rounded the Sun. So 1982i is also 1986 III, as Comet Halley was the third comet to round the Sun in 1986.

Observing new and returning comets can be fun and challenging, for instead of looking at something stationary, you are following an object that moves from night to night. Positions of bright comets are given in astronomy magazines or on computer network bulletins. Make sure that the one you are planning to look for is bright enough to see. If you are a suburban observer, for instance, the comet should be 7th magnitude or brighter. Fainter comets are hard to see against the glow of city skies.

A COMET CRASHES INTO JUPITER

The crash of a comet into a planet is a rare event, but that is what is going to occur with Periodic Comet Shoemaker–Levy 9 and Jupiter in mid-1994. A comet collides with a planet as large as Jupiter, on average, every thousand years or so. Most of these collisions would have taken place unseen by earthly eyes, for the small size of the comet would have made it invisible before impact.

Not so with S–L 9. The comet did us the courtesy of skimming over Jupiter's cloud tops in July of 1992, a near-collision that split the comet into a chain of fragments and caused it to brighten a hundredfold. The comet was bright enough to be observed, and its course toward Jupiter plotted with accuracy.

During the third week of July 1994, most of the telescopes on Earth will be pointed at the comet and Jupiter as they meet for the second and final time.

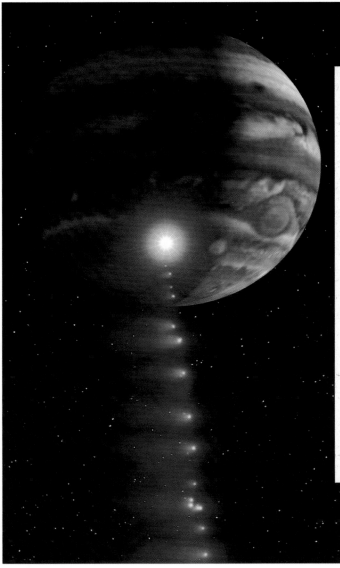

COLLISION! *An artist's conception of S–L 9's fragments (with tails pointing away from the Sun) colliding with Jupiter, each releasing on impact far more energy than a nuclear explosion.*

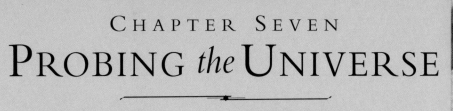

CHAPTER SEVEN

PROBING *the* UNIVERSE

*The universe beckons us to explore beyond our
familiar planet. How have we responded?
What mysteries have we solved,
and what continues to baffle us?*

THE BEGINNING *and* END *of the* UNIVERSE

The "big questions" about the universe are the hardest to answer, but recent years have produced remarkable observations to match some of the amazing theories.

THE GRAND DESIGN *God, the "Great Architect" (above), in an illumination from a thirteenth-century French bible*

Most astronomers now accept that the universe can be broadly explained in terms of the Big Bang theory. This theory says that all matter and space were once compressed into a state of incredibly high temperature and pressure, from which the universe has been expanding ever since.

With the help of computers we can hypothesize that, just a fraction of a second after the Big Bang, the universe was a hot, seething mass of radiation and exotic particles. As it expanded, it cooled and more familiar sorts of matter formed. Eventually hydrogen and helium were produced, which are the main constituents of the universe today.

MICROWAVE BACKGROUND RADIATION

In 1965, Arno Penzias and Robert Wilson of Bell Telephone Labs were tracking sources of static in the receiver system of a radio telescope. Even after they had tightened the telescope's connections and removed some nesting pigeons from the antenna, there was one source of static that they were unable to eliminate. This turned out to be primordial background radiation coming from the entire sky at an extremely cold -455 degrees Fahrenheit (2.7 degrees Celsius above absolute zero)—none other than the echo of the Big Bang itself. The discovery won Penzias and Wilson the 1978 Nobel Prize in Physics.

In April 1992, NASA investigators revealed that their Cosmic Background Explorer (COBE) satellite telescope had detected small variations in the temperature of the background radiation. The COBE data indicate that the universe contained the seeds of its later structure as early as 300,000 years after the Big Bang. From these small irregularities grew the mighty clusters of galaxies, separated by vast voids, which we see today.

THE BIG BANG? *How can we graphically portray the Big Bang? The flash of explosion shown here is clearly not a suitable representation, since we could never be outside the beginning of the entire universe. The Big Bang was an "explosion" of space, not one in space.*

CREATION OF THE UNIVERSE *The Big Bang is widely accepted as having been the origin of all the planets, stars, and galaxies that we see today.*

GRAVITY AND DARK MATTER

What will the universe be like in the distant future? Will it expand forever or will it slow and stop for a split second before starting to contract, ending in a Big Crunch and perhaps another Big Bang?

The key to this question is gravity. If there is enough mass in the universe, gravitational forces will be strong enough to cause the expansion to slow and someday finally stop, then reverse. Otherwise, the expansion will slow, but it will never cease.

Is there enough mass in the universe to stop it expanding? It is clear that more mass is present in the outer parts of our own galaxy than we can see. The stars orbit the galactic center too fast for it to be otherwise. The largest of the nearby clusters of galaxies, the cluster in Virgo, must also

have more mass than we can see or gravity would be insufficient to keep the member galaxies from flying apart.

Where is this dark matter? Hidden in black holes, perhaps, or in the tiny mass that subatomic particles called neutrinos may possess? Somewhere else? We don't know.

In 1993, two groups of astronomers independently reported a new type of invisible object which may form part of the answer—a Massive Compact Halo Object (MACHO). Far away in the outskirts, or halo, of our galaxy, MACHOs are likely to be large, non-shining stars or planets.

Whether or not MACHOs exist, astronomers have not yet detected nearly

enough dark matter to conclude that someday the universe will stop expanding.

EXPANDING FOREVER OR THE BIG CRUNCH?

In an endlessly expanding universe, the galaxies would eventually be filled with ancient stars—black dwarfs that do not shine. After a time vastly more than the current age of the universe, the black dwarfs would decay, leaving nothing but a mist of particles.

But if there is enough mass to halt the expansion, the universe will be in for a very different future. After several billion more years of expansion, there will come an instant when the universe will be at rest. Then, ever so slowly at first, it will begin to collapse on itself, speeding up until, in its last moment, all its matter will coexist as a single point. Then, perhaps, the cycle will start all over again.

Even if there is another Big Bang, the new universe would have no memory of the old. Nothing about our universe would pass through the vortex into the new one. Perhaps even the laws of nature would be different!

RIPPLES IN THE BACK-GROUND OF SPACE *The COBE satellite produced this image of the microwave background radiation—the glow of the Big Bang spread across the sky. The pink and blue ripples represent the origin of the structure we see today as galaxies and clusters of galaxies.*

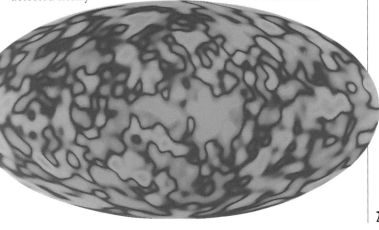

EXPLORING SPACE

*Ever more sophisticated telescopes
and planetary probes have been revealing
the unseen sky at an awe-inspiring rate.*

MARINER 2, *the first US interplanetary
spacecraft, passed within 22,000
miles (35,000 km) of Venus in 1962.*

In the years immediately
following the end of the
Second World War, a
group of American engineers
and scientists at White Sands,
New Mexico, busied them-
selves launching V2 rockets
that had been captured from
the Germans. As the rockets
soared high into the sky, the
engineers learned how they
worked, and the scientists
grew increasingly excited
about the possibilities that
rockets offered for exploration
in space.

However, success came
from another quarter when, on
October 4, 1957, the Soviet
Union stunned the world by
hurling Sputnik 1 into orbit
around the Earth. In response,
the United States began a
program to launch an orbiting
satellite within 90 days. On the
89th day, Explorer 1 made it
into orbit.

In an impassioned speech in
May 1961, President John
Kennedy committed the
United States to landing a man
on the Moon. Despite a fire
that killed three astronauts
early in 1967, an American
finally set foot on the Moon
on July 20, 1969.

BUZZ ALDRIN, *seen in a famous Apollo
11 photo taken by Neil Armstrong (who
can be seen reflected in Aldrin's visor)*

DOG IN SPACE *Sputnik 2 was
launched in 1957, carrying Laika (right),
and instruments to measure the effects
of space flight on the animal.*

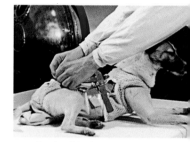

PLANETARY EXPLORATION

Also in the 1960s, a succession
of automated spacecraft began
to explore the planets. NASA's
farsighted program reached its
zenith in the 1970s and '80s,
with Pioneer and Voyager
missions to Jupiter and Saturn,
Mariner 10 visiting Mercury
and Venus, and two Viking
spacecraft landing on Mars.

The highlights of explor-
ation by spacecraft from the
USSR were the Venera series
of missions to Venus in the
1960s and '70s, especially the
photos of the surface made by
Venera 9 and 10 in 1975, and
the Vega 1 and 2 approaches
to Comet Halley in 1986.

Voyager 2 has proved to be
the most successful of all the
automated spacecraft, so far.

Between 1979 and 1989, it visited all four of the gas giant planets, and some of their moons, and provided an enormous amount of information about the Solar System.

Exploration programs are continuing, but political and financial constraints and occasional mission failures have dampened enthusiasm a little.

SPACE OBSERVATORIES

Although attracting less fanfare than the planetary spacecraft, the series of astronomical satellites in Earth's orbit have proved equally revolutionary. They have opened windows on the universe that are inaccessible from the ground and built on work done from small rockets and balloons.

The Einstein X-ray Satellite opened up the X-ray sky between 1978 and 1981. Also launched in 1978, the International Ultraviolet Explorer (IUE) is still surveying the ultraviolet realms. Arguably the most revolutionary of all was the Infrared Astronomy Satellite (IRAS) which surveyed the infrared sky for just six months in 1983, but its data have proved profoundly valuable ever since. New satellites at all these different wavelengths continue to build on the foundations laid by these trailblazers.

THE HUBBLE SPACE TELESCOPE

The most famous, or infamous, satellite observatory is the Hubble Space Telescope (HST). Launched in 1990, its advantage lies not so much in its 90-inch (2.3 m) mirror as in the steadiness of its images as it watches the sky from above the Earth's turbulent atmosphere.

But the HST went aloft with a mirror that focussed poorly. A telescope with such a problem would usually have had its mirror returned to the factory, but that was impossible in this case. It was decided to give the HST "glasses", and these were provided in a package of corrective optics, installed by astronauts in December 1993. Other problems were also corrected and the HST now appears to be performing beautifully.

In the three years between launch and repair, the HST gathered much scientific data and made many discoveries, despite its problems. It is now poised to unlock secrets at the edge of the universe.

We shall not cease from exploration, and the end of all our exploring will be to arrive where we started and know the place for the first time.

Little Gidding,
T.S. ELIOT (1888–1965),
British poet, dramatist, and critic

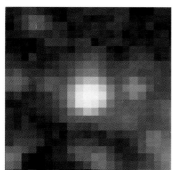

HST VIEWS THE STARS *When the HST was launched, it produced images (center) with much better resolution—much smaller images—than the view from the ground (left) but with an unwanted halo of light. After repair (right), the HST images were much improved, with most light concentrated within the image core as it should be.*

LIFE *in the* UNIVERSE

The possibility of life elsewhere in the universe has always fascinated people. Before long, we may be able to answer the question: "Are we alone?"

We know for sure that there is one planet in our galaxy that supports intelligent life. Is Earth the only such planet, or is it part of some great web of life spanning the galaxy? Intelligent life might exist on many worlds, distance being all that prevents us from making contact.

The search for extra-terrestrial intelligence began in 1960, when radio astronomer Frank Drake started Project Ozma, named after the Queen of Oz in a series of stories by L. Frank Baum. He hoped to find unusual signals from two nearby, Sunlike stars—Epsilon (ε) Eridani and Tau (τ) Ceti—but detected nothing. Attempts continue today, using the most sophisticated techniques.

Radio searches rely on finding signals, either deliberate or accidental. But there might be many life forms out there, ranging from microbes to civilizations, that do not use radio. Perhaps

USS ENTERPRISE *from Star Trek: The Next Generation, presents a view of space travel in a future where contact with other races is routine.*

the only way to detect them is to range as far afield as we can to have a look. We have done this within our Solar System and most astronomers now believe it is unlikely that there is life like that on Earth on any of the other planets in the Sun's domain. Tests by the Viking landers in 1976 at two locations on Mars, the most likely candidate, revealed no signs of life.

ASSORTED ALIENS *Blue men from Mars (above), a pointy-eared Vulcan (below left), and a wrinkled ET (below)*

THE DRAKE EQUATION

Where would intelligent life be most likely to sprout? To provide a starting point, Frank Drake proposed the following formula to calculate N, the number of civilizations in our galaxy with the means of contacting us right now:

$$N = N_* \, f_p \, n_e \, f_l \, f_i \, f_c \, f_L$$

N_* is the number of civilizations in our galaxy

f_p is the fraction of stars with planets

n_e is the number of planets with "suitable" environments

f_l is the fraction of planets where life has arisen

f_i is the fraction of times where life develops "intelligence"

f_c is the fraction of intelligent civilizations capable of calling between the stars

f_L is the fraction of the star's lifetime the civilization survives.

Based on what little we know, we might guess that many of these factors are around 0.1. With around 200 billion stars (N_*) in our galaxy, we conclude that there are 200,000 possible contacts (N) out there.

Some scientists argue that the rise of intelligence might be very rare, so our estimate may be over-optimistic. After all, only one of Earth's uncountable species seems capable of technological communication. Maybe N really does equal 1—us!

IMPACT! *An artist's impression of the moments before (right) and after (below) the impact of an object several miles across. Impacts of this size produced craters many times larger that scar the lunar surface and may be implicated in the extinction of the dinosaurs on Earth.*

IMPACTS UPSET ECOSYSTEMS

During the Earth's early years, the organic material that forms the basis of life may have been deposited by the repeated impacts of comets and asteroids. However, these impacts quite possibly also destroyed life forms that already existed! Life may have had to start repeatedly before it finally took hold.

Such impacts may also have affected the evolution of life. A comet or asteroid at least 6 miles (10 km) wide careened into Earth about 65 million years ago, just off Mexico's Yucatan Peninsula. This impact may have caused, or contributed to, the mass extinction of life forms that occurred at about this time, which included the dinosaurs. If such an impact occurred today, it might well bring our civilization to an end.

LAST DAYS OF THE DINOSAURS

What caused the extinction of the dinosaurs some 65 million years ago? This question has aroused a great deal of controversy and sparked off a multiplicity of theories. The most popular of the likely scenarios goes something like this ... Imagine huge animals peacefully meandering

through lush vegetation—a scene typical of a mild Earth that has lasted millions of years. Then a comet or an asteroid pounds into the planet. Walls of water inundate coastal areas and millions of tons of dust surge upward in a gigantic cloud. Vegetation ignites and the Earth is cloaked in dust and soot. The sky remains black for more than a month, and rain dense with sulfuric acid soaks the land.

If this picture of devastation is correct, the dinosaurs (and countless other animals and plants) could not have survived. With dinosaurs no longer around, other animals, (namely, the early mammals) took over their former niches and were able to evolve to replace them as the dominant animal species on Earth.

DEMISE OF THE DINOSAURS? *Did one or more large meteor impacts bring about the end of the dinosaurs?*

DALEKS, *from the long-running Dr Who TV series, are an example of the imagined threat from life elsewhere, a popular science fiction theme.*

RESOURCES DIRECTORY

Further Reading
Space Places and Organizations
Index and Glossary

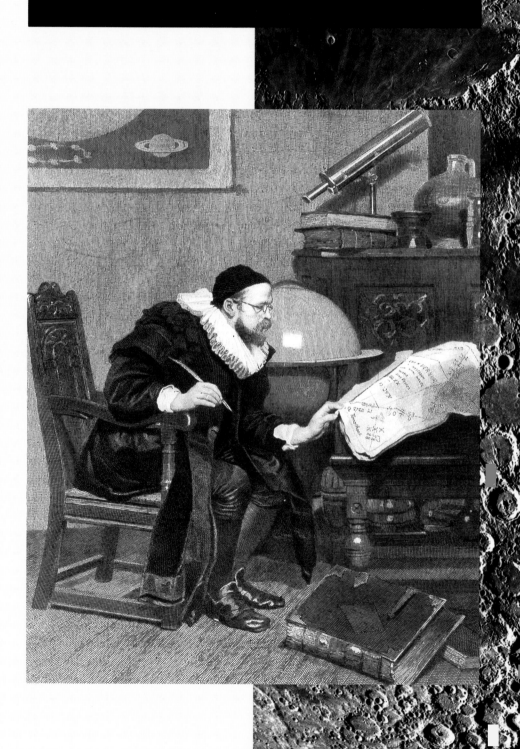

FURTHER READING

Compiled by Robert Burnham

The following publications will guide you deeper into astronomy—both the backyard kind you pursue with binoculars or a telescope and the science you follow from your armchair. The reading levels (given in brackets at the end of each entry) should not be taken too literally and don't hesitate to try one of the more advanced books if the subject interests you. Most of these publications are in print, although few bookstores will carry more than a couple of them. You might like to check out some of the titles at a large library before ordering them.

History of Astronomy

History of Astronomy, by Antonie Pannekoek (Dover, 1989). Although written in 1961, this is still the best general history of astronomy from antiquity up to about 1930. (INTERMEDIATE)

Astronomy of the 20th Century, by Otto Struve and Velta Zebergs (Macmillan, 1962). A really good history of twentieth-century astronomy has yet to be written, but this covers all the major developments up to the 1950s. It is quite readable, in part because Struve knew most of the people he was writing about. (BEGINNER–INTERMEDIATE)

Sidereus Nuncius or The Sidereal Messenger, translated by Albert Van Helden (University of Chicago Press, 1989). Originally published in 1610, Galileo's little book conveys the excitment of seeing the Moon, the Pleiades, the Milky Way, and Jupiter for the first time ever, using the newly-invented telescope. Read this soon after you buy your first telescope—a time when exploring the heavens is as new to you as it was then to Galileo. (INTERMEDIATE)

Mythology and Skylore

Beyond the Blue Horizon, by E.C. Krupp (HarperCollins, 1991). This is a broad, sweeping survey of celestial lore and legends, including those from Native Americans and other indigenous cultures around the world. (BEGINNER–INTERMEDIATE)

Classical Mythology, by Mark P. O. Morford and Robert J. Lenardon (Longman, 1985). While not specifically about astronomy, this is an excellent account of the myths from Greek and Roman antiquity that have become part of our skylore. (INTERMEDIATE)

Star Names, by Richard H. Allen (Dover, 1963). Still in print and widely available, Allen's book gives you a comprehensive look at the history and legends behind star and constellation names as they were understood in 1900. (BEGINNER–INTERMEDIATE)

Star Tales, by Ian Ridpath (Universe Books, 1988). Ridpath retells the legends from classical antiquity that gave rise to the constellations and star names we have inherited as part of Western civilization. (BEGINNER–INTERMEDIATE)

General Astronomy

Astronomy is a colorful and graphic monthly magazine which carries articles on backyard observing and the science of astronomy for the general reader.

Sky & Telescope is also a monthly covering the same fields. It aims its stories more toward the experienced amateur astronomer and is even good reading for professionals.

Exploration of the Universe, by George O. Abell, David Morrison, and Sidney C. Wolff (Saunders College Publishing, 1991). A textbook will give you a detailed survey of contemporary astronomy and makes a good general reference book. This is perhaps the best one written and it is updated every few years. (INTERMEDIATE)

A Concise Dictionary of Astronomy, by Jacqueline Mitton (Oxford University Press, 1991). Just what you need to tell your "periastron" from your "apogee". (BEGINNER)

Stars and Galaxies

Galaxies, by Timothy Ferris (Stewart, Tabori, and Chang, 1982). By far the best armchair tour going—gorgeous color and monochrome photos. The best edition is the first, put out by the Sierra Club in 1980, but even the later printings in reduced format have much of the punch of the original. (BEGINNER)

Stars and Their Spectra, by James Kaler (Cambridge University Press, 1989). Nearly everything astronomers know about stars comes from studying their spectra. Kaler gives you a guided tour of all the different kinds of stars that make up the heavens. (INTERMEDIATE)

Observing Variable Stars, by David H. Levy (Cambridge University Press, 1989). Stars that vary in brightness are among the most fascinating because their changes give astronomers clues as to how they work. Levy explains how anyone with a backyard telescope can follow variables as they wax and wane in brightness. (INTERMEDIATE)

The Milky Way, by Bart Bok (Harvard University Press, 1981). Bok was one of the greats in increasing our knowledge of the Milky Way, and this book will give you a superb picture of the galaxy in which we live. (INTERMEDIATE)

Urban Skywatching

The Urban Astronomer, by Gregory L. Matloff (Wiley, 1991). Few astronomers would choose to observe from a city or the suburbs, but for most of us, that's where we live. Before you dismiss on urban astronomy as hopeless, check out what Matloff has to say. You can see a lot more interesting things than you think. (BEGINNER)

Children and Astronomy

Astronomy for Every Kid, by Janice VanCleave (Wiley, 1991). VanCleave doesn't teach you astronomy as such; instead, she guides you through 100+ simple experiments and science fair projects with everyday objects. These demonstrate principles that govern astronomy and astrophysics, and are also good fun. (BEGINNER)

Observer's Handbooks and Guides

The Sky: A User's Guide, by David H. Levy (Cambridge University Press, 1991). This book gives you an easy introduction to skywatching and amateur astronomy. It offers a fine overview of the subject, setting the stage for you to learn more. (BEGINNER)

Binocular Astronomy, by Craig Crossen (Willmann-Bell, 1992). Most people have binoculars kicking around the house but few think of using them for astronomy. That's a shame and a waste, as Crossen shows in his engaging, folksy way. Those eager to buy a first telescope should read this book before they do. (BEGINNER–INTERMEDIATE)

The Backyard Astronomer's Guide, by Terence Dickinson and Alan Dyer (Camden House, 1991). This will take you the next step beyond *Skywatching*. It covers the entire field of amateur astronomy, with reliable information on what to do, and provides sound advice on equipment. (BEGINNER–INTERMEDIATE)

Telescopes and Accessories

Build Your Own Telescope, by Richard Berry (Scribners, 1985). Despite dozens of commercial instruments being available, many people today still build telescopes them-selves. This book has detailed instructions for five telescopes that are straightforward to build and which will provide excellent performance. (INTERMEDIATE–ADVANCED)

Astrophotography

Astrophotography For the Amateur, by Michael A. Covington (Cambridge University Press, 1991). Short, clearly written, and rich in practical advice, this is the first book to get if you want to try celestial picture-taking with conven-tional camera and film. (BEGINNER–INTERMEDIATE)

Introduction to Astronomical Image Processing, by Richard Berry (Willmann-Bell, 1991). Electronic imaging is where astrophotography is heading. Berry gets you started in processing images on a computer. The book provides both software and sample images you can work with on an IBM PC. (INTERMEDIATE)

Choosing and Using a CCD Camera, by Richard Berry (Willmann-Bell, 1992). Thoroughly practical and filled with detail, this is the book to have if you're wondering what all the fuss is about with CCD (charge-coupled device) cameras and imaging. (INTERMEDIATE–ADVANCED)

Computers

Astronomical Algorithms, by Jean Meeus (Willmann-Bell, 1991). Most amateur astronomers run commercial software on their computers rather than doing their own astronomical

programming, but if you're interested in your own number-crunching, this is an excellent guide. You'll have to write your own code (in Basic, Pascal, or whatever), but this book provides the equations and procedures commonly used in astronomy applications. (INTERMEDIATE–ADVANCED)

Learning Constellations

Discover the Stars, by Richard Berry (Harmony Books, 1987). There's no great trick to it, says Berry. Just find a star figure you can recognize, such as the Big Dipper or Orion, and step off from there. (BEGINNER)

Celestial Lights

Meteors, by Neil Bone (Sky Publishing, 1993). A basic, good guide to get you going watching meteors, plotting their tracks on star maps, and taking photos of them. (INTERMEDIATE)

Meteor Showers, by Gary W. Kronk (Enslow, 1988). Kronk takes you month by month through the year, identifying all major and minor showers. (INTERMEDIATE)

Thunderstones and Shooting Stars, by Robert T. Dodd (Harvard University Press, 1986). A short but detailed and readable explanation of meteorites and their parent bodies, the asteroids. (INTERMEDIATE)

The Meteorite and Tektite Collectors' Handbook, by Philip M. Bagnall (Willmann-Bell, 1991). A handy guide to a growing field of interest among amateur astronomers (and among rock and mineral collectors too). (INTERMEDIATE)

The Aurora Watcher's Handbook, by Neil Davis (University of Alaska Press, 1992). Well illustrated and clearly written, this will appeal to anyone fascinated by the eerie flickering and pulsing of the northern (or southern) lights. Lots of how-to details on observing and photographing them. (INTERMEDIATE)

Majestic Lights, by Robert H. Eather (American Geophysical Union, 1980). Written by a geophysicist, this large-format book on the history of auroral studies showcases hundreds of illustrations. (BEGINNER)

Light and Color in the Outdoors, by M.G.J. Minnaert (Springer-Verlag, 1993). This is the book to have if you want to understand what makes rainbows, moonbows, sundogs, mirages, and a host of other phenomena you'll see in the atmosphere. (BEGINNER–INTERMEDIATE)

Star Atlases

Sky Atlas 2000.0, by Wil Tirion (Cambridge University Press, 1981). This is the best all-round star atlas and it covers the whole sky. Buy it as a beginner and you'll still be using it years and years from now. (BEGINNER)

Uranometria 2000.0, by Wil Tirion, Barry Rappaport, and George Lovi (Willmann-Bell, 1987 and 1988). This is the atlas every advanced amateur astronomer should own. There is also a matching Deep-Sky Field Guide to Uranometria 2000.0 (1992), which lists the basic data on every non-stellar object plotted on the Uranometria charts. (INTERMEDIATE–ADVANCED)

Star Catalogues

Sky Catalogue 2000.0, edited by Alan Hirshfeld and Roger W. Sinnott (Cambridge University Press, 1981 and 1985). Sky Cat lists basic data such as positions, brightnesses, sizes, distances, and so on. Volume 1 covers stars of 8th magnitude or brighter; Volume 2 covers double, multiple, and variable stars, together with galaxies, star clusters, and nebulae. While Sky Cat was designed to accompany Sky Atlas 2000.0, it is highly useful on its own. (INTERMEDIATE)

Burnham's Celestial Handbook, by Robert Burnham, Jr. (Dover, 1977-1978). In these three volumes, a former Lowell Observatory astronomer pulls together a vast array of information on all types of celestial objects. More than 15 years old, this is still the first reference book many amateurs reach for. (BEGINNER–INTERMEDIATE)

The Messier Album, by John H. Mallas and Evered Kreimer (Cambridge University Press, 1979). Charles Messier's list is renowned for including the best and prettiest deep-sky objects visible in small telescopes. Mallas and Kreimer give a short description of each object, together with a sketch and a photo. (BEGINNER–INTERMEDIATE)

The Solar System

The Grand Tour, by Ron Miller and William K. Hartmann (Workman Publishing, 1993). In wonderful color paintings and simple descriptions, Hartmann and Miller take you on a tour of the Solar System. The book was recently updated to include information from the final Voyager encounters with Uranus and Neptune. (BEGINNER)

The New Solar System, edited by J. Kelly Beatty and Andrew Chaikin (Cambridge University Press, 1990). Now in its third edition, this contains chapters about every Solar System topic. The level of presentation is about the same as Scientific American magazine. (INTERMEDIATE–ADVANCED)

Planetary Landscapes, by Ronald Greeley (Chapman & Hall, 1994). Revised to include the results from Voyager 2's flyby of Neptune and its moon Triton, this is the best guidebook to the surface geology of the Moon and the planets in the Solar System. (INTERMEDIATE)

Stardust to Planets, by Harry Y. McSween, Jr. (St. Martin's Press, 1993). Engagingly written, this is a good survey of the Solar System for armchair astronomers. (BEGINNER)

Moons and Planets, by William K. Hartmann (Wadsworth, 1992). If you want to know how the Solar System works, this is the best introductory textbook on the subject and is quite readable. (INTERMEDIATE–ADVANCED)

Introduction to Observing and Photographing the Solar System, by Thomas A. Dobbins, Donald C. Parker, and Charles F. Capen (Willmann-Bell, 1988). This is a practical guidebook to exploring the Sun's family with your backyard telescope. (INTERMEDIATE)

The Restless Sun, by Donat G. Wentzel (Smithsonian Institution Press, 1989). Besides telling the story of how astronomers think the Sun works, Wentzel also explores questions such as a possible relationship between sunspot cycles and Earth's climate. (INTERMEDIATE)

Observing the Sun, by Peter O. Taylor (Cambridge University Press, 1991). One of the joys of solar observing is that you don't need dark skies. Taylor explains how you can safely view the Sun with any telescope, providing you take a few simple precautions. (INTERMEDIATE)

The Moon, by Michael T. Kitt (Kalmbach Publishing, 1992). This is the best guide to lunar observing with a backyard telescope. Kitt gives you a tour of the lunar surface (night by night for a month) and explains what you're seeing, using geological information from the Apollo missions. (BEGINNER–INTERMEDIATE)

Atlas of the Moon, by Antonin Ruekl (Kalmbach Publishing, 1990). A detailed atlas that depicts the craters, mountains, and plains of the near side of the Moon in 76 beautiful, precisely drawn maps. The book includes chapters on lunar feature names and space missions and has a small portfolio of excellent lunar photos. (INTERMEDIATE)

Mercury: the Elusive Planet, by Robert G. Strom (Smithsonian Institution Press, 1987). Although we've seen only half of it, Mercury is an unusual world that superficially resembles the Moon but whose interior is more like Earth's. (INTERMEDIATE)

The Evening Star: Venus Observed, by Henry S. F. Cooper (Farrar Straus Giroux, 1993). Cooper takes you behind the scenes with the scientists of the Magellan project, which sent a spacecraft to map the surface of Venus (a world as large as Earth), using radar to penetrate the planet's opaque clouds. (BEGINNER)

Venus: The Geological Story, by Peter Cattermole (John Hopkins University Press, 1994). More technical than *The Evening Star,* this book has all the latest results from the Magellan mission and is thoroughly illustrated. (INTERMEDIATE)

The Earth's Dynamic Systems, by W. Kenneth Hamblin (Macmillan, 1989). This textbook for non-science college students is the best layperson's guide to the forces and processes operating on the world around you. Very well illustrated. (INTERMEDIATE)

Mars, by Peter Cattermole (Chapman and Hall, 1992). A good deal has been learned about the Red Planet in the last 20 years of space missions. This is an excellent up-to-date overview of Mars—its volcanos, craters, deserts, atmosphere, and ice caps. (INTERMEDIATE–ADVANCED)

The Planet Jupiter, by B. M. Peek (Faber and Faber, 1981). You'll not find a more detailed guide to the ever-changing telescopic appearance of Jupiter. Dated somewhat by the Voyager flybys, the book is still of unsurpassed usefulness to backyard observers. (INTERMEDIATE–ADVANCED)

Voyage to Jupiter, David Morrison and Jane Samz (NASA, 1980). This NASA publication details the flybys of Jupiter by the Voyager 1 and 2 spacecraft. The book blends "you-are-there" mission history with the initial scientific results and contains lots of photographs. (BEGINNER)

The Planet Saturn, by A. F. O'D. Alexander (Dover, 1980). This and Alexander's book on Uranus, listed below, are historical surveys of their respective planets. Both books provide a rich source of pre-Voyager information and are wonderful archives to delve into on a cloudy night. (INTERMEDIATE–ADVANCED)

Voyages to Saturn, David Morrison (NASA, 1982). This does for the Saturn flybys what *Voyage to Jupiter* did for Jupiter. Once again, lots of photos and a must-read text for anyone interested in Saturn. (BEGINNER)

The Planet Uranus, by A. F. O'D. Alexander (Faber and Faber, 1965). This is the same sort of work as Alexander's book on Saturn, described above. It covers the "prehistory" of Uranus—the period up to the Voyager 2 encounter. (INTERMEDIATE–ADVANCED)

Uranus, by Ellis Miner (Ellis Horwood, 1990). Following on from Alexander's book (above), Miner's brings the story up-to-date through the Voyager 2 encounter, with all its scientific findings presented clearly. (INTERMEDIATE)

Planets X and Pluto, by William G. Hoyt (University of Arizona Press, 1980). This retells the wonderful story of how wrong measurements, erroneous assumptions, and off-track guesswork led astronomers to find the planet Pluto. (BEGINNER)

Out of the Darkness; the Planet Pluto, by Clyde W. Tombaugh and Patrick Moore (Stackpole Books, 1980). Clyde Tombaugh was a 24-year-old ex-amateur astronomer when he discovered Pluto. This is his story of the events leading up to the discovery and what was learned about this small, distant world up to the 50th anniversary of the discovery. (BEGINNER)

Introduction to Asteroids, by Clifford Cunningham (Willmann-Bell, 1988). Nearly 20,000 asteroids are known, but very few have been studied in any detail. Cunningham looks at everything from the numbers of asteroids that have been discovered and their types, to the chances that Earth will be hit by one again soon. (INTERMEDIATE–ADVANCED)

Rendezvous in Space, by John C. Brandt and Robert D. Chapman (W.H. Freeman, 1992). Brandt and Chapman tell you what astronomers have learned about comets using the latest research, including the much-studied return of Halley's Comet in 1986. The book contains computer program listings (in Pascal and Basic) to calculate the positions of a comet, given its orbit. (INTERMEDIATE)

Cosmology

Lonely Hearts of the Cosmos, by Dennis Overbye (HarperCollins, 1991). Who would have thought that the field of cosmology would be so rich in personalities? Overbye's wonderful book lets you hang out with the professional astronomers who work with the world's biggest telescopes in search of the bounds of the universe. After reading this, you'll never look at astronomers the same way again. (BEGINNER)

The Big Bang, by Joseph Silk (W.H. Freeman, 1989). Whether you see it as scientific dogma or simply the best available model for the universe, the Big Bang definitely stirs people's opinions. Silk shows clearly how the model emerged from nearly a century's research, and also why it has proved so robust. (INTERMEDIATE–ADVANCED)

Cosmology, by Edward Harrison (Cambridge University Press, 1981). This book is a comprehensive guide to the major concepts in cosmology. Harison deals with some major misconceptions and presents a range of historical material. (INTERMEDIATE–ADVANCED)

Space Exploration

Journey into Space, by Bruce Murray (Norton, 1989). Subtitled "the first thirty years of space exploration", this is much more than just a history. Murray, a planetary scientist at Caltech, also takes the time to make a passionate case for why people should explore space. (BEGINNER)

Space Age, by William J. Walter (Random House, 1992). This is the companion to a 6-part PBS series, yet it can stand alone as a good, illustrated history of the space era. It explores how space perspectives have altered the way humans look at themselves and their home planet. (BEGINNER)

The Illustrated Encyclopedia of Space Technology, by Kenneth Gatland (Orion Books, 1989). This is the perfect resource book for hardware fanatics. Gatland uses photos and detailed cutaway drawings to show you the insides of spacecraft such as Mercury, Vostok, Soyuz, and the Shuttle. (INTERMEDIATE)

Search for Extraterrestrial Intelligence (SETI)

Is Anyone Out There?, by Frank Drake and Dava Sobel (Delacorte Press, 1992). Well, that is The Big Question, isn't it? While Drake and Sobel don't have any definitive answers, they outline what SETI astronomers are doing and how they hope to improve their search methods. Drake is one of the field's pioneers. (BEGINNER)

The Cosmic Water Hole, by Emmanuel Davoust (MIT Press, 1991). This book looks more broadly at the question of what life is. Starting with the searches for life within the Solar System, Davoust then examines the forms in which it

might exist elsewhere and how we should try to hunt for it. The book has an extensive bibliography. (BEGINNER–INTERMEDIATE)

Periodicals

U.S.

Astronomy is a colorful and graphic monthly magazine devoted to presenting both the science of astronomy and the joys of backyard observing to the amateur astronomer.

CCD Astronomy is a quarterly devoted entirely to electronic imaging, a new development attracting much interest among astrophotographers.

Mercury is a bimonthly sent to members of the Astronomical Society of the Pacific. It contains articles on astronomy written for the lay reader.

The Planetary Report covers the latest developments in Solar System exploration, and is sent to members of The Planetary Society.

Sky & Telescope is a monthly covering the entire subject of astronomy. Its stories are aimed toward experienced amateur astronomers and even the professional.

The Strolling Astronomer is a bimonthly published by the Association of Lunar and Planetary Observers. The articles appeal primarily to the dedicated amateur astronomer with a strong interest in viewing planets.

CANADA

Journal of the Royal Astronomical Society of Canada is a bimonthly for amateur and professional astronomers.

BRITAIN

Astronomy Now is a popular-level monthly which combines articles about the science with how-to features for observers.

Journal of the British Astronomical Association is a bimonthly publication and the "journal of record" for the BAA. Most stories appeal to the intermediate to advanced amateur.

AUSTRALIA

Sky and Space (formerly *Southern Astronomy*) is a colorfully illustrated monthly that presents the science of astronomy and how-to articles with a Southern Hemisphere emphasis.

Computer Software

Expert Astronomer (IBM and Macintosh) is an inexpensive and basic "planetarium" program. It shows you the sky from any given location at a given date and time.

Dance of the Planets (IBM) is a tour de force that has superb graphics and helps you visualize the Solar System from anywhere within it, on any date.

Voyager II (Macintosh) is a slick and powerful planetarium program with over 50,000 stars and 4,200 deep-sky objects.

(Check in current astronomy magazines for updates and reports of new software in this rapidly growing field.)

Besides running software on their PCs, many amateur astronomers are using modems to contact dial-in bulletin boards and networks, and many have begun to explore the vast resources of the Internet. Here is a list of toll-free numbers for some commercial on-line services:

American Online (800) 827 6364
CompuServe (800) 848 8199
Delphi (800) 695 4005
Genie (800) 638 9636
Prodigy (800) 776 3449

SPACE PLACES *and* ORGANIZATIONS

Space places

Australia Telescope National Facility, (HQ) P.O. Box 76, Epping, NSW 2121, Australia. Operates both the Parkes Radio Observatory and the Australia Telescope Radio Interferometer. Good visitors centers.

Anglo–Australian Telescope, (HQ) P.O. Box 296, Epping, NSW 2121, Australia. Located in the Warrumbungles National Park. Home of the 150 inch (3.9m) telescope, the largest in Australia. Excellent visitors center.

Dominion Radio Astrophysical Observatory, Windlake Road, Penticton, British Columbia, Canada V2A 6K3. Operates a 85ft (26m) radio disk, plus a 2,000 ft (600m) aperture-synthesis array. Offers self-guided tours.

Dominion Astrophysical Observatory, West Saanich Road, Victoria, British Columbia, Canada V8X 4M6. Has exhibits, tours, and a 6ft (1.85m) reflector for visitors.

Jodrell Bank Observatory, Macclesfield, Cheshire SK11 9DL, England. Home of the giant 250-ft radio telescope, the largest fully steerable dish in the world.

Kitt Peak National Observatory, PO Box 26732, Tucson, Arizona 85726, USA. Operates a 160 inch (4 m) and many other telescopes, including solar ones.

Lick Observatory, PO Box 55, Mount Hamilton, California 95140, USA. Has a 120 inch (3 m) reflector and a 36 inch (900 mm) refractor.

Lowell Observatory, 1400 W. Mars Hill Road, Flagstaff, Arizona 86001, USA. Has a 24 inch (600 mm) Clark refractor used by Percival Lowell for Mars studies,and the 13 inch (320 mm) telescope used to discover Pluto.

Mauna Kea Observatory, Hilo, Hawaii, USA. Houses the world's largest telescope (the 30 foot [10 m] Keck telescope).

Mount Wilson Observatory, 740 Holladay Road, Pasadena, California 91106, USA. Operates a 100 inch (2.5 m) reflector.

National Air and Space Museum, The Mall, Washington, DC 20560, USA. The best museum in the world for anyone interested in space and aviation.

Old Royal Observatory, Greenwich, England. Of great historical interest and is located on the world's prime meridian.

Palomar Observatory, California Institute of Technology Pasadena, California 91125, USA. Operates the 200 inch (5 m) Hale telescope.

Very Large Array (VLA), PO Box O, Socorro, New Mexico 87801, USA. Operates a radio telescope made of 27 radio dishes with arms 7 miles (11 km) long. Also headquarters for the Very Long Baseline Array, a multi-antenna telescope with dishes placed from Hawaii to Puerto Rico.

Yerkes Observatory, 373 W. Geneva Street, Williams Bay, Wisconsin 53191, USA. This is home to the world's largest refractor telescope, with a 40 inch (1 m) lens.

Organizations

American Association of Variable Star Observers (AAVSO), 25 Birch Street, Cambridge, Massachussets 02138, USA. A group of (mainly) amateur astronomers who keep thousands of variable stars under watch.

Association of Lunar and Planetary Observers (ALPO), PO Box 143, Heber Springs, Arizona 72543, USA. An amateur group that carries out planetary patrol observations.

Astronomical Society of the Pacific (ASP), 390 Ashton Avenue, San Francisco, California 94112, USA. A national organization of amateur and professional astronomers.

British Astronomical Association (BAA), Burlington House, Piccadilly, London W1V 9AG, England. The national society for amateur astronomers.

International Dark Sky Association, 3545 N. Stewart Street, Tucson, Arizona 85716, USA. Dedicated to preserving dark skies for skywatchers.

The Planetary Society, 65 N. Catalina Avenue, Pasadena, California 91106. A space-advocacy group interested in Solar System exploration by both unmanned spacecraft and human expeditions.

Royal Astronomical Society of Canada (RASC), 136 Dupont Street, Toronto, Ontario, Canada M5R 1V2. The national organization for professional and amateur astronomers; has local "centers" (clubs) in major cities across Canada.

Royal Astronomical Society of New Zealand (RASNZ), PO Box 3181, Wellington, New Zealand. The national society and a reporting center for Australian variable star observations.

To find your nearest astronomy club or planetarium, look in the listings periodically published in astronomy magazines or ask a local telescope retailer.

INDEX *and* GLOSSARY

I n this combined index and glossary, bold page numbers
indicate the main reference, and italics indicate illustrations
and photographs.

44, **45**
SX Phoenicis 199
Syrtis Major 247, *247*

T

T Columbae 158
T Coronae Borealis 161
T Pyxidis 203
Tarantula Nebula *34*, 45, **167**, *167*
Tau (τ) Canis Majoris 147
Tau Ceti 27, 91
Taurids 96
Taurus 54, 109, 117, 120, **214–15**, *214*, *215*
Taygete 215
techniques and tools 54–77 *see also* observing and observation techniques
telescopes 21, *21*, 22, *22*, 23, 36, *52–3*, **62–7**, *62*, *63*, *64*, *65*, *66*, *67*, 70, 72, 75, **76–7**, *76*, 77, 107, 238, **269**
Telescopium 131, **216**, *216*
Telrad A sighting device for telescopes which projects a bull's-eye on the sky. **67**
terminator 239
Thales 18–19, 222
Theta (θ) Apodis 135
Theta (θ) Arae *138*
Theta (θ) Carinae 150
Theta (θ) Eridani 170
Theta (θ) Leonis 181
Theta (θ) Orionis 37
Theta (θ) Scorpi 138
Thuban 168
tides **92**, *92*, 93, **239**
time 86–7, 104
time zones 86–7, *87*
Titan 250–1
Titania 252, *252*, 253
Titius, Johann 258
Titius-Bode Law 258
Tombaugh, Clyde 22, 23, 51, **58**, 172, 178, **257**, *257*
Tombaugh's Star 162
transit The instant when a celestial object crosses the meridian. 241
Trapezium 195
Triangulum 119, **217**, *217*
Triangulum Australe 131, **218**, *218*
Trifid Nebula *43*, 45, **267**
tripods 60, *60*, 64, 68
Triton 254, *255*
Trojans *258–9*, **259**
Tropic of Cancer 144
Tropic of Capricorn 149

Tucana 123, 127, 131, **219**, *219*
Tucanae, **47** 39,*39*, **219**, *219*
TV Corvi 162
Tycho *236*, 238

U

U Arae 139
U Sagittae 205
ultraviolet (UV) The portion of the spectrum with wavelengths just shorter than the bluest light visible. 23, **77**, 91, 269
umbra The dark inner part of an eclipse shadow. Also, the dark central part of a sunspot. *95*, 233
Umbriel 252, *252*, 253
Unidentified Flying Objects (UFOs) **99**, *99*
Universal Time (UT) 87
Universe 18–21, *19*, 50–1, **266–7**, *267*, **270–1**
Uranus 23, *71*, 172, **252–3**, *252*, *253*
Ursa Major 16, 38, *102*, 103, *103*, 112, 141, 145, 182, **220–1**, *220*, *221*, 222
Ursa Minor *29*, *102*, 108, 112, 141, 145, 221, **222**, *222*
Ursids 96

V

V Hydrae 176
V Sagittae 205
Vallis Marineris 247
Van Maanen's Star 201
variable star Any star, the brightness of which appears to change, with periods ranging from minutes to years. **40–1**, *40*, *41*, 72, **74**, **184**, 187
Vega 90, 91, 106, 112, 114, 115, 124, 126, 164, 186
Veil Nebula 165, *165*
Vela 121, 125, 131, 150, **223**, *223*
Vela supernova remnant *45*
Venus 23, 55, 69, 93, 95, **242–3**, *242*, *243*, 258, 268
vernal equinox *84*, 86
Very Large Array (VLA) 23, *23*, 77
Vesta 259
Virgo 51, 103, *103*, 113, 122, 124, 159, **224–5**, *224*, *225*, 267
Virgo Cluster 224, *225*
Virgo's Diamond 111
visual (V) magnitudes 91
Volans 127, **226**, *226*
volcanoes 242, 243, 245, 247, 248
Vulpecula 227, *227*

VW Hydri 178
VZ Camelopardalis 143

W

Webb, Thomas 166
West (1975a), Comet 261
Whirlpool Galaxy 48, 71, **145**, *145*
white dwarf The small and hot, but intrinsically faint, remnant left when a red giant star loses its outer layers as a planetary nebula. 31, 34, *37*, 41
Wild Duck Cluster 39, **211**, *211*
Wilson, Robert 22, 23, 266
winter solstice 84, 85, *85*
Wolf 359 27, 91
"Wolf-Rayet" star 223

X

X-rays 23, 77, *233*, 269

Y

Y Canum Venaticorum 145
Yerkes, Charles 76
Yerkes Observatory 23, 76

Z

Z Camelopardalis 143
Z Chamaeleontis 156
zenith The point on the celestial sphere directly overhead. 82, *81*, 82, *82*, 105
Zeta (ζ) Cephei 154
Zeta (ζ) Draconis 168
Zeta (ζ) Geminorum 40
Zeta (ζ) Orionis 195
Zeta (ζ) Piscium 201
Zeta (ζ) Puppis 203
Zeta (ζ) Scorpii 209
Zeta (ζ) Ursa Majoris *see* Mizar
zodiac The 12 constellations straddling the ecliptic through which the Sun, Moon, and planets appear to move during the year. **17**, *17*, 84
Zuben El Genubi 183
Zuben Eschamali 88, 183

ACKNOWLEDGEMENTS

The publishers wish to thank in particular the following people for their assistance in the production of this book: Jason Forbes, Eve Gordon, Selena Hand, Greg Hassall, Tim Hunter, Mary O'Byrne, Peter Hingley (Royal Astronomical Society), Larry Lebofsky, Nancy Lebofsky, Jean Mueller, Jonathan Nally (Editor, *Sky & Space*), Jim Scotti, Eugene Shoemaker, Carolyn Shoemaker, Oliver Strewe, Lesa Ward, David Warren, and the staff at the Anglo-Australian Observatory.

PICTURE CREDITS
(t = top, b = bottom, l = left, r = right, c = center)

CAPTIONS
Page 1: A fourteenth-century Italian bronze of an astronomer.
Page 2: Tycho Brahe, Ptolemy, St Augustine, Copernicus,
Galileo, and Andreas Cellarius with The Muse of Astronomy.
Page 3: The Sun-centered universe and the signs of the zodiac.
Pages 4–5: The moon may look larger when near the horizon.
Pages 6–7: Portrait of the Milky Way by Jon Lomberg.
Pages 8–9: A detail from Van Gogh's *The Starry Night* (1889).
Pages 10–11: A portion of Johann Bode's atlas *Uranographia*
(1801). Detail: Capricornus the sea goat.

Pages 12–13: The School of Athens by Raphael (1483–1520).
Pages 24–5: The famous Horsehead Nebula in Orion.
Pages 78–9: A spectacular aurora over Wasilla, Alaska.
Pages 52–3: Preparing for another night of skywatching.
Pages 100–1: Constellation patterns depicted in a fresco in the
Palazzo Farnese, Italy.
Pages 228–9: The relative sizes of the planets and the Sun.
Page 264–5: Repairing the Hubble Space Telescope, 1994.
Page 272: A portion of the brilliant Milky Way in Sagittarius.